VISUAL MECHANICS

 BEAMS & STRESS STATES

Gregory R. Miller
University of Washington

Stephen C. Cooper
University of Washington

PWS PUBLISHING COMPANY

I(T)P *An International Thomson Publishing Company*

Boston • Albany • Bonn • Cincinnati • Detroit • London • Madrid
Melbourne • Mexico City • New York • Pacific Grove • Paris
San Francisco • Singapore • Tokyo • Toronto • Washington

PWS PUBLISHING COMPANY
20 Park Plaza, Boston, MA 02116-4324

Windows 95™ is a registered trademark of Microsoft, Inc.
Macintosh™ and MacOS™ are registered trademarks of Apple Computer, Inc.
Java™ is a registered trademark of Sun Computer Systems, Inc.
Dr. Beam™ and Dr. Stress™ are trademarks of Dr. Software, LLC.

Contact Information

PWS Publishing Company
20 Park Plaza
Boston, MA 02116-4324

Dr. Software, LLC
10516 41st Place NE
Seattle, WA 98125

Developmental Editor: Mary Thomas-Stone
Technology Editor: Leslie Bondaryk
Marketing Manager: Nathan Wilbur
Production Coordinator: Elise Kaiser
Manufacturing Buyer: Andrew Christensen
Cover Designer: Diane Levy
Cover Image: David Sailors © 1995 The Stock Market, Inc. Image is of construction on the central dome and roof structure at Washington National Airport; Cesar Pelli, architect.
Cover Printer: Phoenix Color Corporation
Text Printer/Binder: Malloy Lithographing

Printed in the United States of America.
97 98 99 00 01 — 10 9 8 7 6 5 4 3 2 1

ISBN 0-53495-587-8

Contents

▶ **GERE & TIMOSHENKO CHAPTER 5**

▶ GERE & TIMOSHENKO CHAPTER 6

► **GERE & TIMOSHENKO CHAPTER 9**

STRESS STATES   116

Preface

The material in this manual and the associated software are intended to complement the study of two of the principal topics in mechanics of materials: (i) beam bending; and (ii) analysis of stress states. The software has been designed to support and encourage exploration and visualization of the phenomena and mathematical models that are covered in the text, and it can help accelerate your progress in developing the kind of understanding that comes with the experience of seeing and interpreting solutions to many different problems. The programs are simple to use, but they also are quite powerful — as your understanding of the material develops, it is likely that you will find the programs continually useful.

The manual itself has two main components: (i) general worksheets; and (ii) touchpoints associated with specific text discussions and examples found in Gere and Timoshenko's *Mechanics of Materials, Fourth Edition.* The worksheets and touchpoints provide structured activities intended to help you make your own observations and discoveries concerning the basic behavior of beams, further consequences of fundamental theories, and generalizations of special cases. (There were limitations in the number and variety of exercises that could be included in the workbook—visit the Visual Mechanics web site at http://www.pws.com/pws/vismech.html for additional exercises and other interesting things to do with the software.)

Programs like Dr. Beam and Dr. Stress reduce to seconds analyses that can take hours to do by hand, and they can generate literally hundreds of solutions per minute. Don't be fooled, though: engineering is not simply a matter of generating correct numbers to particular problems any more than the practice of medicine is simply a matter of taking good x-rays. Word processors make it easier to write, but they do not make it any easier to be a good writer—so with engineering software and engineers. For any topic in engineering science there is a clear need to learn how to solve problems, but one must also develop an understanding of what solutions mean, what sorts of analyses are appropriate and useful to perform, and how to interpret the accuracy and engineering significance of whatever results are obtained. This caveat having been noted, you are encouraged to play around with these tools, and have some fun as you learn. The exercises presented here work well for groups as well as for individual study—by discussing the various posed questions with your colleagues, you are likely to learn more.

Getting Started with the Software

- **Installation**

  The CD supplied with this workbook contains the software and documentation necessary to support the exercises and activities presented here. Installation instructions for each platform (Macintosh, Windows 95, or Unix) can be found on the CD itself.

- **Running the Programs**

  Once installed, the Macintosh and Windows 95 versions of Dr. Beam and Dr. Stress can be started in any of the usual ways, including double-clicking the program icon or any associated file icon, or using drag-and-drop. The Unix version will require a command line entry to invoke the Java runtime environment (the Unix version should actually be executable on any Java platform).

  The first time the programs are run, it will be necessary to enter a registration number to enable the programs. The appropriate registration numbers are given below:

  Dr. Stress: 041-397-526-464

  Dr. Beam: 073-911-265-067

- **Internet**

  Be sure to check out the Visual Mechanics web site for FAQs, updates, additional exercises, discussions, feedback and whatever else we or you can think of. The url is http:// www.pws.com/pws/vismech.html.

- **Getting Help**

  Both Dr. Beam and Dr. Stress have online help available in the menus. In addition, the CD contains refernce documents for both programs. Any questions which are not answered by these sources should be directed to the Visual Mechanics web site: http:// www.pws.com/pws/vismech.html.

BEAMS

Worksheet 1
(G&T Ch 4 Introduction)

AN INTRODUCTION TO BEAM BENDING

1 ■ OVERVIEW AND CONTEXT

Of all the means for transferring loads, bending is actually one of the least efficient (think about how you naturally try to break things like sticks by bending them across your knee—ever try pulling one apart in tension?). However, the general utility and easy fabrication of flat surfaces and straight geometries makes bending elements very common. Even for elements not necessarily intended to act as beams (e.g., truss members), secondary bending effects are likely to arise. Because bending is both relatively inefficient and nearly ubiquitous in applications, it frequently is a governing factor in design. You will need to learn the fundamental theory underlying beam analysis and methods for generating analytical solutions, but you also need to develop a qualitative sense of how beams behave and what to expect from an analysis. This worksheet can help in regard to this latter objective.

This worksheet is structured around a set of fundamental questions and associated activities intended to help you address these questions. The presentation here is relatively elementary: most of the issues introduced here will be revisited in more detail later.

2 ■ HOW DO BEAMS WORK?

Try this experiment: pick up a standard 8-foot long 2 x 4 with one hand by lifting it at its center. The average person will have no trouble getting the board up in the air. Now repeat the exercise by lifting the board by one of its ends. In this case it will be extremely difficult to get the entire board off the ground and hold it horizontal to the ground. What's going on here? The object being lifted and the person doing the lifting have not changed, yet the result is quite different.

■ Things to Do

1. To see what is happening, consider the free-body diagrams shown below:

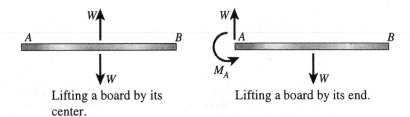

Lifting a board by its center.

Lifting a board by its end.

What is the magnitude of the moment your hand needs to supply in the center-lifting case? _____

What is the magnitude of the moment your hand needs to supply in the end-lifting case? (remember the beam is 8-ft long) _____

2. To supply the moment in the end-lifting case, the moment arm your hand has to work with is about the size of the beam's depth, which is a dimension significantly less than the beam's length.

    Assuming a 3.5-in moment arm for your hand, how big do the couple forces have to be to supply the supporting moment?_____lbs

3. For the end-lifting case, compare the magnitude of the couple forces your hand needs to apply to the magnitude of the weight of the board.

    The ratio is ❏ greater than 1 ❏ less than 1.

The board's weight in effect has substantial leverage, and so your hand must work at a mechanical disadvantage.

4. Now imagine you are strong enough to support the board by its end. Think about the material in the beam adjacent to your hand; this material must itself generate the same kind of moment as your hand, and like your hand must work at a significant mechanical disadvantage because it, too, only has the beam depth to work with. At a simple level, this mechanical

disadvantage is the crux of why beams are relatively inefficient in carrying loads, and why moments in a beam turn out to be the primary concern in regards to strength: moments arising from loads on beams tend to result in highly magnified internal forces (stresses).

▶ **Observation** | Beams function by transforming applied loads into internal shears and moments. The moments in particular can lead to highly amplified internal forces.

3 ■ WHAT ARE THE GENERAL RESPONSE CHARACTERISTICS OF A BEAM UNDER LOADING?

It will take a while to answer this question fully — in this worksheet we will just get started. In particular, we will use Dr. Beam to make direct observations in light of the following more specific questions.

3.1 What quantities do we use to characterize beam response?

Before worrying about what a beam's response will be, we need to define what we mean by "response" in a quantitative sense, and that is what we will try to accomplish here.

■ Things to Do

1. Open Dr. Beam if necessary, select the point load tool ⬇️, and load the startup beam by clicking anywhere on the beam and dragging up and down and back and forth.

■ Comment

As you load the beam, you will see dynamically updated plots of the three main quantities that characterize a beam's response to load: (i) the *displaced shape* of the beam, which is drawn on the beam itself; (ii) the *shear diagram*, plotted below the beam and labeled with a *V*; and (iii) the *moment diagram*, plotted last and labeled with an *M*. The displaced shape provides an exaggerated depiction of what the beam looks like when deformed, the shear diagram shows how the internal shear forces vary along the beam's length, and the moment diagram similarly characterizes the variation of the bending moment as a function of location along the beam.

In addition to displacements, shears, and moments in a beam, there is one other quantity that can play a role in characterizing a beam's behavior. To see what it is, do the following:

2. Choose the **Plots** command from the **Options** menu and click the remaining unselected check box. Click **OK**, and you will now see four plots. See if you can figure out what the new plot means physically, and then turn this plot back off using the **Plots** command again.

▶ **Observation**

The primary response characteristics of a beam are moment, shear, and displacement, which are functions of location, and which can be characterized using simple plots. The displacement slope is also of interest in certain applications.

3.2 How do the fundamental response quantities relate to the board lifting exercise described above?

To gain some additional feel for what these response quantities mean physically, it is useful to consider the simple board lifting exercise in terms of these quantities.

■ **Things to Do**

1. Continuing with Dr. Beam, choose the **Clear** command in the **Edit** menu to reset the beam. This will leave you with an unloaded simple beam.

2. To model a case analogous to that in which you picked the beam up in the middle, choose the fixed support tool ⬛ and click near the beam's center (if you want to be precise, double click on the new support and use the dialog box to set the location exactly).

3. Now use the selection tool (the arrow at the top of the palette) to remove the end supports one at a time by clicking on them and typing the **delete** key. The beam is now supported at its center only.

4. Now add a load to model the beam's weight. Use the point load tool ⬛ to place a point load at the center support (again you can be precise by double clicking the load and setting the location and magnitude exactly).

Do you see any non-zero values on any of the plots? ❑ yes ❑ no

5. To model the case in which your hand is not at the beam's center, use the select tool ⬛ to click on the center support, and then press the right or left cursor keys on your keyboard to move the support toward either end (if you are in a hurry, hold down the shift key, too).

6. Observe the moment and shear diagrams as the support (analogous to your hand holding the board) moves toward the beam's end. (Depending on how big a point load you put on the beam you may need to rescale the moment plot to keep it on the screen. Rescaling is accomplished by clicking and holding the up and down arrows to the right of the plots. ⬛)

   Do you see any non-zero values on the plots now? ❑ yes ❑ no

   As the support moves from the center the support moment ❑ increases ❑ decreases ❑ stays the same.

   As the support moves from the center the support shear ❑ increases ❑ decreases ❑ stays the same.

▪ Comment

The rapidly increasing moment you see with increasing span shows again why moments tend to be the most critical factor for strength considerations in designing beams. Let's look at this quantitatively.

7. Choose the **Show Values** command in the **Options** menu. If you look at the feedback panel at the bottom of the window after turning on **Show Values**, you will see that as the mouse moves within the beam span, numerical values from the plots are displayed corresponding to the cursor's location.

8. Use this feature to determine the moment and shear values required at the support (the pseudo hand) when the support is at the end of the beam by locating the mouse at the support, and reading off the desired values from the feedback panel.

   Shear =_____

   Moment =_____

   (don't forget your units)

■ Comments

You should verify these values using simple statics—even with Dr. Beam, it's good to get a second opinion. If you have time, try putting in realistic values for the board we have been discussing and see how strong you would need to be to hold up the board.

While this analysis will correctly indicate the support reactions your hand must supply when holding on at the beam's end, there is a fundamental difference between this analysis and the actual situation as far as the beam itself is concerned: the board's weight is distributed uniformly along its length rather than being concentrated at its center. Let's model the actual case by using a distributed load.

9. Remove the point load by selecting it and typing the **delete** key. Apply a distributed load along the entire length of the beam using the distributed load tool ▦. You can do this either by double clicking and entering the appropriate values in the dialog box, or by clicking and dragging horizontally as you apply the load. Again, depending on the magnitude of the load you apply, you may need to adjust the plot scales using the scale arrows to the right of each plot. Note that the magnitude of the load can be adjusted by clicking and dragging vertically after the load has been applied.

    If the magnitude of the distributed load is adjusted such that the total load applied is the same as in the previous point load case, the moment and shear at the support are ❏ greater than ❏ equal to ❏ less than the point load case.

10. Now drag the support back to the beam's center, and think about the physical situation being modeled:

    What are the reactions *you* must apply to the beam when lifting it at its center?

    Vertical reaction = _____

    Moment reaction = _____

■ Comment

At first glance, these values may not appear to agree with the Dr. Beam plots. The plotted moment in particular is not zero at the center. Remember, though, that the plots show the moment and shear *in* the beam, not the external reactions. Even though it is easy for you to lift the beam by its center, the beam must still keep itself straight, which requires significant moment resistance (think about what happens if you pick up a piece of rope at its center). Now let's review how to relate internal and external shears and moments.

11. Evaluate the jump in the shear at the center support. This can be accomplished either by placing labels on each side of the jump, or by moving the mouse across the jump while reading the values off the feedback pane at the bottom of the window. (To place labels, use the label tool 🔳 to click at the desired location on the desired plot. Double clicking enables accurate location and more precise value readout.)

Jump in shear = _____ k

Total applied load = _____ k

These two quantities ❏ are ❏ are not equal.

12. Now use the right or left arrow key to move the center support toward one of the ends again, and carefully observe the moment diagram.

The jump in the moment across the support is a ❏ maximum ❏ minimum when the support is at the beam's center.

The jump in the shear ❏ does ❏ does not depend on the support location.

■ Comment

The jumps or discontinuities that arise in these moment and shear diagrams represent the external moment that the support (your hand, in the current context) must apply to the beam to maintain equilibrium. Again, this is an easy case to verify using simple statics, and you should do so for a particular support location. Note how the large moment required for the end support case matches your experience with picking up a beam at its end.

By setting the distributed load magnitude to correspond to the board's weight/unit length (double click on the load to set its end magnitudes), you can determine the moment capacity of your hand experimentally. Determine how far from the center you can hold the board and still pick it up, and then set the support location in Dr. Beam to match this distance (or put the appropriate values into your simple statics calculations). Calculate the corresponding moment jump at the support, and there you have it: your hand's moment capacity.

3.3 How are a beam's response characteristics related to load location and magnitude?

Having gained a basic idea about how a beam's response is quantified, it's reasonable to consider how this response depends on basic load, support, and cross-section properties. This is the focus of the remaining questions in this section.

■ Things to Do

1. Use the **Clear** command in the **Edit** menu to return to the standard case of a beam with simple end supports. Use the point load tool ⊞ to add a new point load to the beam.

2. Experiment by moving the load around and adjusting its magnitude using the mouse. Based on your observations, determine whether each of the following statements is true or false. (Note: if you find yourself creating unwanted additional loads, you can remove them by clicking on them and typing the delete key on your keyboard.)

   The maximum moment always occurs at the middle of a beam, regardless of load location. ❏ true ❏ false

   The maximum displacement always occurs at the middle of a beam, regardless of load location. ❏ true ❏ false

   Loads placed near supports cause bigger moments than loading away from supports. ❏ true ❏ false

   Loads placed near supports cause bigger shears than loading away from supports. ❏ true ❏ false

   Loads placed near supports cause bigger displacements than loading away from supports. ❏ true ❏ false

   Maximum displacements always occur at points of maximum load. ❏ true ❏ false

3.4 How are a beam's response characteristics related to load type?

To answer this question we will consider the basic shape and continuity characteristics of the various response plots for different load types.

■ Things to Do

1. Make sure you still have a single point load on the beam, and describe each plot you see as either piecewise constant, piecewise linear, or smoothly curved by checking the appropriate boxes below (Note: if necessary, use the scaling buttons to magnify the plot shapes.)

   Displaced shape:

   ❏ piecewise constant ❏ piecewise linear ❏ smoothly curved

   Shear diagram:

   ❏ piecewise constant ❏ piecewise linear ❏ smoothly curved

   Moment diagram:

   ❏ piecewise constant ❏ piecewise linear ❏ smoothly curved

2. Add two more point loads to the beam at new locations and answer the previous question again (if your plots start to overlap, use the scale buttons to the right of each plot to adjust the scales to a suitable size):

   Displaced shape:

   ❏ piecewise constant ❏ piecewise linear ❏ smoothly curved

   Shear diagram:

   ❏ piecewise constant ❏ piecewise linear ❏ smoothly curved

   Moment diagram:

   ❏ piecewise constant ❏ piecewise linear ❏ smoothly curved

3. Now remove all the loads by using the **Clear** command in the **Edit** menu, and apply a distributed load by selecting the distributed load tool 🔲 and clicking anywhere on the beam. For the case of a distributed load, how do the plots turn out?

   Displaced shape:

   ❏ piecewise constant ❏ piecewise linear ❏ smoothly curved

   Shear diagram:

   ❏ piecewise constant ❏ piecewise linear ❏ smoothly curved

   Moment diagram:

   ❏ piecewise constant ❏ piecewise linear ❏ smoothly curved

4. In addition to observing the general shape of the various plots, it is useful to consider their continuity. Use the **Clear** command and the point load tool ⊞P to return to the single point load case and characterize each plot's continuity:

   Displaced shape:

   ❑ discontinuous ❑ kinked ❑ smoothly continuous

   Shear diagram:

   ❑ discontinuous ❑ kinked ❑ smoothly continuous

   Moment diagram:

   ❑ discontinuous ❑ kinked ❑ smoothly continuous

5. Delete the point load, and replace it with a concentrated moment by clicking and dragging with the concentrated moment tool ⟳. Answer the continuity question again.

   Displaced shape:

   ❑ discontinuous ❑ kinked ❑ smoothly continuous

   Shear diagram:

   ❑ discontinuous ❑ kinked ❑ smoothly continuous

   Moment diagram:

   ❑ discontinuous ❑ kinked ❑ smoothly continuous

▶ **Observation** | The general shape and continuity characteristics of the moment and shear diagrams depend on the type of loading applied.

3.5 How are a beam's response characteristics related to its support conditions?

The manner in which a beam is supported is another consideration for analysis and design. The influence of different support conditions is examined here.

■ **Things to Do**

1. Clear the beam again and put on a single point load. Make a note of the value of the moment at the left end of the beam.

   Original value: ❑ negative ❑ zero ❑ positive

2. Now choose the fixed support tool ⬚ and click on the beam's left support. The simple support is replaced by a fixed support — how did the value of the moment at the left end change?

Final value: ☐ negative ☐ zero ☐ positive

3. Check off the appropriate shape observations below for the new support condition, and compare your observations to the original case you considered above in Section 3.4.1.

Displaced shape:

☐ piecewise constant ☐ piecewise linear ☐ smoothly curved

Shear diagram:

☐ piecewise constant ☐ piecewise linear ☐ smoothly curved

Moment diagram:

☐ piecewise constant ☐ piecewise linear ☐ smoothly curved

4. By alternately using the fixed and simple support ⬚ tools (or using **undo/redo** in the **Edit** menu) you can toggle back and forth between these two cases—see what further observations you can make about how the shear, moment, and displacements change as the support conditions change.

5. One often thinks of supports as preventing displacement at precisely located points along a beam. In reality, all supports have some degree of "give" (*compliance*). Also, they can settle over time, or they can be placed imprecisely. In some cases, effects of this kind can cause substantial deformation and loading in the beam. To see an example of this, make the beam's left support fixed, and remove all the loads. All the plots should be zero. Now click on the left support with the fixed support tool ⬚ and drag it up and down to model settlement or misalignment. Note the resulting moments, shears, and displacements.

6. To quantify these observations, select the label tool ⬚ and click at the origin of each plot one at a time. This will create value labels that provide numerical feedback at discrete locations (as in the case of loads and supports, you can place a label precisely by double clicking on it and entering the desired location. The location dialog box also gives the label's value with more significant figures than the onscreen representation). Now as you drag the support vertically, you can monitor the actual values at the beam's left end. Determine if the induced moments vary linearly with (i.e., are proportional to) the applied support displacement. (You can set precise support displacements by double clicking on them with the selection tool.)

The moment variation is: ❏ linear ❏ nonlinear.

7. Without removing your labels, replace the fixed support one more time with a simple support, and repeat the vertical dragging of the left support.

In this case, settlement/misalignment ❏ does ❏ does not induce bending effects.

▶ **Observations** Support type and location influence the response of beams. Support settlement can induce loads and deformations in beams.

3.6 How are a beam's response characteristics interrelated?

There are important fundamental relations between the various response characteristics of a beam, most of which can be surmised by direct observation.

■ **Things to Do**

1. Clear the beam and place a distributed load over the entire span using the distributed load tool ⊞. (This is most easily accomplished by clicking at the left end and then dragging to the right end. Once the load is created you can adjust its magnitude by clicking and dragging vertically anywhere in the loaded interval.)

2. Use the **Plots...** command in the **Options** menu to turn on the slope plot, and then fill in the appropriate letters to complete the following statements. (Hint: look at how maxima and zeros line up, and consider the degree of curvature of the plots. Remember to use the scaling buttons to magnify the plots if you can't discern the shapes easily.)

Answer list: a. rotation b. moment c. shear d. displacement

The load diagram looks like the derivative of the _____ diagram.

The shear diagram looks like the derivative of the _____ diagram.

The moment diagram looks like the derivative of the _____ diagram.

The rotation diagram looks like the derivative of the _____ diagram.

As demonstrated mathematically in Gere and Timoshenko (see Sections 4.2 and 9.2), the plots are indeed related by derivatives. This illustrates the underlying basis of beam analysis: solving differential equations.

▶ **Observation** | Beam analysis can be viewed as solving (i.e., integrating) differential equations.

3.7 How are a beam's response characteristics related to cross-sectional properties?

As part of the board exercise considered earlier, it was observed that the relatively small moment arm represented by a beam's depth leads to high internal stresses in a beam. It should follow that by making a beam deeper, one can increase its efficiency. One measure of this efficiency is a beam's relative stiffness. By bending beams with the cross-sections shown below about their horizontal and vertical axes, you can get a qualitative feel for the relation between depth and stiffness.

As you may recall from physics, the quantity that represents the distribution of material about an axis is the moment of inertia. For a cross-section, the moment of inertia collapses to the second moment of area, I. The quantity I provides a measure of how much material in a cross-section is far away from the bending axis, where it can be most effective in resisting bending.

We can use Dr. Beam to get a quantitative measure of how a beam's behavior varies with I.

■ **Things to Do**

1. Clear the beam using the **Clear** command in the **Edit** menu (you can turn off the rotation plot, too, using the **Plots...** command in the **Options** menu) and load it using the point load tool ⊞. Place labels on the shear, moment, and displacement diagrams somewhere near the load application point. Record the values below:

    Shear =_____

    Moment = _____

    Displacement =_____

2. Use the **Beam Properties...** command in the **Info** menu to open a dialog box that will allow you to set various beam properties. In particular, increase the default value for I in the dialog box by a factor of 2, click the **Apply** button, and watch how the curves change. Record the new label values below:

   Shear =_____

   Moment = _____

   Displacement = _____

3. Compute the ratios of original and current values and compare these to the factor of 2 we applied to I:

   Shear Ratio = _____

   Moment Ratio = _____

   Displacement Ratio =_____

4. Predict what will happen if you double I again, and then check your prediction. See if you agree with the following statement:

   The displacements in a beam depend linearly on the second moment of area of the cross-section, I. ❑ I agree ❑ I disagree.

▶ **Observation** | All else being equal, a beam's stiffness can be increased by using a cross-section with a larger I.

3.8   How are a beam's response characteristics related to its material properties?

■ **Things to Do**

1. Open the **Beam Properties...** dialog again, and repeat the above activity, except this time vary the material stiffness, E. You can use the values you just recorded above as the baseline values—this time halve rather than double E, click **Apply**, and record the new values below:

   Shear = _____

   Moment = _____

   Displacement = _____

2. Compute the ratios again, and compare this to the factor applied to E:

Shear ratio = _____

Moment ratio = _____

Displacement ratio = _____

3. Try another prediction/observation experiment, and see if you agree with the following statement:

The displacements in a beam depend linearly on the modulus of elasticity, E.
❏ I agree ❏ I disagree.

▶ **Observation** | Using a stiffer material in a beam can help reduce displacements, but this will not influence moments and shears in general.

3.9   How are a beam's response characteristics related to geometry?

The principal geometric property of a beam other than its cross-section profile is the length of its span(s). Here we consider the effect of varying span length on the response.

■ **Things to Do**

1. Open the **Beam Properties...** dialog once again, and this time double the beam's length. Watch what happens, and record your observations below:

Shear = _____

Moment = _____

Displacement = _____

2. Compute the following ratios between the original and length-doubled cases.

Shear Ratio = _____

Moment Ratio = _____

Displacement Ratio = _____

■ Comments

In this case, the story is quite different than when we varied the cross-section and material properties. In particular, the displacement varies like the square of the length, while the moment varies linearly. These are not general observations, though—different loadings and support configurations will lead to different relations. If you experiment with different loads and geometries, however, you can make the following generalization.

▶ **Observation** | The displacements in a beam are very sensitive to beam length. Moments in a beam are increased by increasing beam length, sometimes dramatically.

4 ■ WHAT ARE SOME LIMITS OF LINEAR, 1-D BEAM ANALYSIS?

Whether you do your beam analysis by hand or with a computer program, there are limitations to the underlying model itself. An analytical tool or theory generally will not allow you to explore its own limitations—one must call on experiments or other theories for this purpose. Engineers have the responsibility to know the limits of whatever tool, theory, or method they are using. Pay careful attention as you read the derivations in your text, and note each assumption as it is made.

One of the main limitations of the predictive power of linear, 1-D beam analysis can be observed easily. Take the board that you used earlier and set it up to span between two books with the tall dimension oriented vertically. Load the beam with books or weights until it fails, i.e., collapses. Note that the failure mode is not associated with basic bending, per se, but rather arises from the beam tipping over. The issue of lateral stability (i.e., tipping or rolling over) is very important in beam design, but simple beam analysis does not include this phenomenon. Had you determined the board's capacity using simple beam concepts alone, you would have been grossly in error.

▶ **Observation** | Beam design can not be based on linear, 1-D analysis alone.

5 ■ SUMMARY

This worksheet has provided an overview of elementary concepts concerning beam behavior. As you develop an increased understanding of beam theory, you may find it helpful to revisit these exercises.

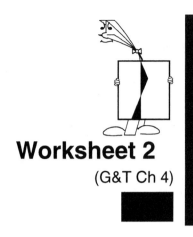

Worksheet 2

(G&T Ch 4)

UNDERSTANDING MOMENT DIAGRAMS

1 ■ OVERVIEW AND CONTEXT

In this worksheet the focus is on understanding and interpreting moment diagrams qualitatively as well as quantitatively. By learning how moment diagrams are supposed to look for different loadings and support conditions, and by understanding the qualitative behavior of beams, you can train yourself to sketch reasonable moment diagrams prior to doing any analysis. Not only does this help guide you as you work through quantitative analyses, it also equips you to view beams more like an engineer does, mentally translating load and support conditions into distributions of moments.

2 ■ ARE THERE HELPFUL ANALOGIES FOR MOMENT DIAGRAMS?

A common device in engineering is to relate an abstract, new, or unfamiliar system to a familiar or easily visualized system by means of an analogy. One very useful analogy for understanding moment diagrams comes out of looking more closely at the load-shear-moment relations:

$$\frac{dV}{dx} = -q \,; \; \frac{dM}{dx} = V$$

We can eliminate the shear, V, from the above relations by taking the derivative of the second equation with respect to x, and then substituting for V in the first equation. This gives us the following equation relating the moment to the applied load directly.

$$\frac{d^2M}{dx^2} = -q$$

If you think back to when you learned your physics, at some point you probably studied wave propagation in stretched strings (e.g., guitar strings, clotheslines, etc.). As part of that study you would have derived the following relation between the tension in the string, T, the displacement of the string, v, and the load applied to the string, q:

$$T\frac{d^2v}{dx^2} = q$$

By comparing this equation to the above relation for moments and loads, we can make the following useful observation:

▶ **Observation** | Moment diagrams in a beam are analogous to the displaced shape of a clothesline subjected to the same loading.

Let's see if this works, and if so, what it tells us.

■ **Things to Do**

1. Open the file **Worksheet2a** and apply a point load anywhere in the span. Move it around and adjust its magnitude by dragging the mouse.

   The moment diagram for a given load looks like the deformed shape of a stretched string carrying the same load, except the moment diagram is ❏ upside down ❏ backwards.

   Note: in many parts of the world it is common to use a sign convention for moments such that the string analogy is even more direct.

2. Apply multiple loads and move them around. Try two loads with one pointing up and one pointing down.

   It still looks like a loaded string. ❏ true ❏ false

3. Replace the point loads with a distributed load by selecting each point load, typing the **delete** key, and then clicking once with the distributed load tool ⊞ to add the new load. Use the mouse to drag the load around and change its magnitude. By clicking and dragging vertically near the end of the load you can create a linearly varying distributed load, as well.

   It still looks like a string. ❏ true ❏ false

19

4. For each case above note how the shape is just what your intuition about stretched strings would predict. Look in particular for regions where the string would be straight (i.e., linear moment variation), and where kinks would occur. Answer each of the following, returning to the load cases above, if necessary.

In regions of beams between loads (i.e., in unloaded regions), the moment diagram will always be: ❏ straight ❏ flat ❏ curved ❏ kinked.

In regions of beams under distributed loads, the moment diagram will always be: ❏ straight ❏ flat ❏ curved ❏ kinked.

At points where concentrated loads exist, the moment diagram will always be: ❏ straight ❏ flat ❏ curved ❏ kinked.

At the end points of a distributed load, will the moment diagram have a kink? ❏ yes ❏ no ❏ maybe

The kinking of the moment diagram associated with a concentrated load will be directed ❏ toward ❏ away from the applied load.

The curvature of the moment diagram under a distributed load will be ❏ toward ❏ away from the applied load.

■ Comment

Note that each of the above observations follows directly from a simple analogy with stretched strings. You should now be able to sketch reasonable moment diagrams for most any simply-supported beam without doing any calculations. This analogy can only be pushed so far, though, as the following illustrates.

5. Use the **Clear** command in the **Edit** menu to return the beam to its unloaded state, and use the point load tool ⬇ᴾ once again to put a single load on the beam. Now use the fixed support tool ⬛ to replace the left pin support by clicking on it.

What feature(s) of the plot has(have) changed?

❏ presence of kink

❏ location of kink

❏ direction of kink (peak up/peak down)

❏ curvature of diagram (straight/curved)

❏ boundary values of plot

■ Comment

For general boundary conditions the simple string analogy does not tell the whole story, but the general observations concerning the local shape and curvature of the moment diagram still apply. This follows from the fact that the differential equations relating loads to moments and loads to string displacements are analogous, but the boundary conditions only match in the simply-supported case. The practical result of this is that you can tell everything about the general shape of a moment diagram within a particular region just by thinking about how a string would behave locally under the same loading, but we need additional observations to handle general boundary conditions.

3 ■ WHAT ARE THE CONDITIONS FOR SYMMETRIC MOMENT DIAGRAMS?

Symmetry is a very useful visual cue to check moment diagrams, and it is also helpful in making the mathematical analysis of many problems simpler. The key is to know when symmetry applies.

■ Things to Do

1. Open the files **Worksheet2b-f**, and for each file experiment with different support conditions and locations and load magnitudes and locations. Determine which of the following ingredients are necessary for a symmetric moment diagram:

    ❑ Symmetrically located supports

    ❑ Symmetrically typed supports

    ❑ Loads with symmetric values

    ❑ Loads with symmetric locations

    ❑ Pin supports only

    ❑ Concentrated loads only

2. Here's a question you can use to stump your friends and possibly instructors, as well: Is it possible to have a symmetric moment diagram with a non-symmetric displacement? If you use the **Plots...** command in the **Options** menu to turn on displacements for each of the files considered in step 1, you might be led to conclude that the answer is no. Take a look at **Worksheet2g**, though, for a counterexample.

By noting the symmetry properties of the geometry and loading of a beam, you can tell a priori whether the moment diagram will be symmetric or not.

4 ■ HOW DO SUPPORT/BOUNDARY CONDITIONS EFFECT MOMENT DIAGRAMS?

We have already seen that general boundary conditions foul up the simple string analogy that works so well for simply-supported beams. Nevertheless general boundary conditions are a part of life and must be dealt with accordingly. Coming to an understanding of how general supports influence a moment diagram largely boils down to understanding when moments must vanish at a support or boundary, and when they normally will not vanish.

■ Things to Do

1. Open the Dr. Beam file **Worksheet2h** using the **Open** command in the **File** menu, and see if you can find any loading that causes non-zero moments at either of the simple supports.

   I was ❑ successful ❑ unsuccessful.

2. Change the left support to a fixed support by clicking on it with the fixed support tool ⊡. See if you can find a load combination that makes the moment vanish at this fixed support.

   I was ❑ successful ❑ unsuccessful.

3. Add an additional pin support near the center of the beam by clicking there with the pin support tool ⊡. See if you can find a load combination that will make the moment at the pin support vanish.

   I was ❑ successful ❑ unsuccessful.

4. Remove the left support by clicking on it with either the selection tool ⊡ or the fixed support tool ⊡ and typing the **delete** key. See if you can find a loading to cause non-zero moment at the now unsupported left end of the beam.

   I was ❑ successful ❑ unsuccessful.

5. Return the fixed support to the left end of the beam, and then use the internal hinge tool $\boxed{\,\circ\!\!\!-\,}$ to insert a hinge in the beam between the left and center supports. Try to find a load combination that causes non-zero moment at the hinge.

I was ❏ successful ❏ unsuccessful.

6. Based on your observations above indicate which of the following statements are valid:

❏ At a pinned end of a beam, the moment must vanish (unless a concentrated moment is applied right at the support).

❏ At a fixed end of a beam, the moment generally will not vanish except for very special loading conditions (see e.g., **Worksheet2i**)

❏ At an internal simple support, the moment must vanish.

❏ At a free end of a beam, the moment must vanish (unless a moment load is applied directly at the beam's end).

❏ At an internal hinge, the moment generally will not vanish except for very special loading conditions.

▶ **Observation** | By considering carefully the support conditions in a beam, it is possible to locate points of zero moment without analysis.

5 ■ WHAT CAN THE DISPLACED SHAPE TELL ME ABOUT THE MOMENT DIAGRAM?

You may not learn how to calculate displaced shapes in beams until later, but you probably already have a reasonably well-developed intuition about what a deformed beam should look like for a given load. By learning how moment diagrams relate to displaced shapes, you will be able to leverage your pre-existing intuition into a useful companion for moment diagram recognition and construction.

■ Things to Do

1. Open the Dr. Beam file **Worksheet2j** using the **Open** command in the **File** menu, and use the point load tool $\boxed{\,\downarrow\,}$ to apply a downwards load somewhere in the beam's span. Note that the displacement plot is activated in this file, and so the displaced shape of the beam is plotted along with the moment diagram.

For the downwards load the displaced shape is concave ❑ up ❑ down.

The figure below shows a simple illustration of the deformation in a bent beam:

For the displaced shape currently on the screen, the compression in the beam is on the ❑ top ❑ bottom portion of the beam.

The moment diagram for this case is plotted ❑ above (i.e., on top of) ❑ below (i.e., on the bottom of) the zero-moment axis.

2. Replace the left support of the beam with a fixed support, and observe the corresponding displaced shape and moment diagram.

When the moment plots negative (i.e., on the bottom of the plot) the beam is concave ❑ up ❑ down and the compression is in the ❑ top ❑ bottom portion of the beam.

When the moment plots positive (i.e., on the top of the plot) the beam is concave ❑ up ❑ down and the compression is in the ❑ top ❑ bottom portion of the beam.

3. Try additional load and support conditions, and see if you agree with the following statement:

The moment diagram always follows the compression side of the beam. ❑ I agree ❑ I disagree.

■ **Comment**

In the touchpoint Beams 3 this same link is made between displaced shapes and moment diagrams, and some additional practice exercises are provided.

6 ■ HOW DO I PUT THE PIECES TOGETHER TO SKETCH MOMENT DIAGRAMS?

It is possible to construct reasonable moment diagram sketches completely using the four basic principles contained in the exercises above:

(i) consider the shape issues associated with the string analogy (i.e., presence and locations of kinks, curvature of the moment diagram, etc.);

(ii) look for symmetries in the geometry and loading;

(iii) use any a priori knowledge of vanishing moments due to support and boundary conditions; and

(iv) sketch a displaced shape to determine where the moment is positive and negative.

With a little practice, this can all become second nature, and Dr. Beam is an excellent companion for practicing.

■ Things to Do

1. Open the file **Worksheet2k**. Note that all the plots have been turned off, so you can work out your own moment diagram sketch without any help from Dr. Beam. Once you are done, you can check how you did. We will use the steps outlined above to help generate a prediction.

   (i) Shape issues (think about an upside down, stretched string to answer these):

   For the region of the beam under the distributed load, the moment diagram will be ❏ curved ❏ straight.

   If it is curved, it will be concave ❏ up ❏ down.

   For the region of the beam between the distributed load and the point load, the moment diagram will be ❏ curved ❏ straight.

   If it is curved in this region, it will be concave ❏ up ❏ down.

   For the region of the beam between the point load and the right support, the moment diagram will be ❏ curved ❏ straight.

   If it is curved in this region, it will be concave ❏ up ❏ down.

   Will there be any kinks in the moment diagram? ❏ yes ❏ no

   If so, where will this/these kink(s) be located? _____

   _____

   If they exist, this/these kink(s) point ❏ up ❏ down.

   (ii) Symmetries

   The moment diagram for this problem ❏ will ❏ will not be symmetric.

   (iii) Boundary and support conditions (a priori knowledge of zero moments)

At the left support the moment will be ❑ zero ❑ non-zero.

At the right support the moment will be ❑ zero ❑ non-zero.

(iv) Displaced shape

Draw a sketch of the displaced shape on the figure below. Recall that the fixed support forces the slope of the curve to be zero at the left end of the beam.

Based on your sketch, identify regions where the compression (and the moment diagram) is on the top of the beam, and where the compression is on the bottom of the beam. Note that at any crossover points the moment will be zero. (If you are not confident in your sketch, you can get some help from Dr. Beam by turning on the displacement plot using the **Plots...** command in the **Options** menu. Don't turn on the moment plot yet, though!)

If you now try to construct a moment plot incorporating all the information we have determined so far, you will find that only one curve will work, and this will be the correct one (in a qualitative sense—you still need to do calculations to get numbers and relative magnitudes). Go ahead and try it on the following figure, and then have Dr. Beam show you the correct result by turning on the moment plot.

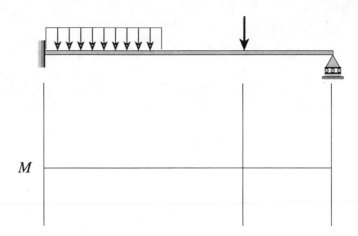

■ **Comment**

In many cases this process will not lead to a unique solution because the relative magnitudes of the various applied loads sometimes can be crucial. Even in these ambiguous cases the process outlined above still will lead to a small number of possible candidate curves that will be similar to one another.

Your textbook has a large supply of beam problems on which you can practice this process. Sit down with your book, Dr. Beam, and some paper and see how good you can get at constructing reasonable moment diagrams by inspection.

7 ■ HOW DO MOMENT DIAGRAM CONCEPTS RELATE TO STRUCTURAL FORM?

As you develop an ability to visualize and sketch moment diagrams for given load and support conditions, it is also useful to understand what information a moment diagram can provide in regards to structural form. As we will see shortly, the shape of a moment diagram can provide significant guidance in regards to defining an efficient structural geometry to accommodate the given loadings for the specified support conditions.

■ **Things to do**

1. Open the file **Worksheet2l**. A distributed load has been applied to the beam shown, but the magnitude of the load is zero.

2. Use the distributed load tool ⊞ to increase the load's magnitude by clicking and dragging down anywhere within the loaded interval. Look at the resulting moment diagram and answer the following:

   I ❏ have ❏ have not seen bridges shaped like this.

3. Either imagine the moment diagram turned upside down, or use the distributed load tool ⊞ to drag the load up.

   What type of bridge has this characteristic shape? _____

4. Open the file **Worksheet2m**. Two point loads have been applied to a simply supported beam, and once again their magnitudes are zero.

5. The loads have been grouped, so if you click and drag down on either one using the point load tool ⊞, they will both increase together. Once you have increased the loads, answer the following:

   I ❏ have ❏ have not seen truss bridges with a similar outline.

6. Open the file **Worksheet2n**. A distributed load with zero magnitude has been applied to a cantilever. Imagine that the cantilever is actually a tall structure, with the ground corresponding to the fixed support, and you are looking at the world lying on your side. The distributed load can be thought of as a simple model of a wind load.

7. Make the wind blow using the distributed load tool ⊞ to drag the load up and down (i.e., back and forth relative to the structure). Note that the moment envelope has been turned on, so the plot keeps a record of the maximum value at each point. If you keep the envelope symmetric, it will help you do the following:

   Name a famous structure whose outline is similar to the red moment envelope you see. _____

■ Comment

We have seen that a moment diagram is analogous to a deformed string. A string can only carry its loads by means of tension, and tension is a very efficient way to carry loads. The upside down string represented by a moment diagram corresponds to a structure that can only carry loads by means of compression. By modeling a structure after these shapes, it is possible to transfer a bulk of the loads by means of simple tension and compression, which is an efficient way to use material. This is particularly true in truss-like structures, but even beams can be made more efficient by varying their depth to be consistent with the moment distribution. (Note that in the case of beams this does not lead to a true optimum—see the Fully Stressed Beams section in Gere and Timoshenko, Section 5.8).

▶ **Observation** | Moment diagrams represent efficient structural forms.

8. See what other structural forms you can generate by considering moment diagrams or moment envelopes for different types of loads, load histories, and supports. Identify 3-5 real structures whose shapes you can relate to moment diagrams

▶ **Summary Observation** | In addition to being at the heart of beam design, moment diagrams provide a useful engineering perspective on structural behavior. Becoming proficient at qualitative beam analysis is not difficult to achieve using the four basic principles outlined above. Such proficiency provides an important complement to quantitative analysis skills.

Worksheet 3
(G&T Section 9.4)

HOW DR. BEAM WORKS

1 ■ OVERVIEW AND CONTEXT

The approach to beam analysis upon which Dr. Beam is based is straightforward, and builds directly on the concepts used for determining displacements by integration. In short, Dr. Beam uses a divide and conquer strategy, breaking any particular beam into a series of simple elements whose individual response characteristics are known, and then mathematically piecing together the system response as an appropriate sum of each element's individual response. There is a useful physical interpretation of what this process of analysis accomplishes, and Dr. Beam can provide some self-illumination in this regard.

Here we first present the mathematical background of how Dr. Beam generates solutions, followed by some hands-on exercises demonstrating the physical viewpoint. You may prefer to look at the exercises first, and then return to the mathematics.

2 ■ THE MATHEMATICS OF DR. BEAM

The figure below shows the fundamental building block used by Dr. Beam for its analysis. The figure depicts a portion of a beam located between two points, x_i and x_j. This segment of the beam is carrying a linearly varying load, and its local boundary conditions are given in terms of the displacements and slopes at each of its ends. Note that the load parameters w_i and w_j can be zero, so the formulation includes the case of an unloaded element.

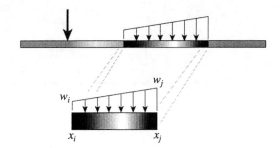

A segment of a beam

The governing equation and boundary conditions for this segment of the beam can be stated as follows:

$$EIv'''' = q(x)$$

Boundary conditions:

$$v(x_i) = v_i$$
$$v'(x_i) = \theta_i$$
$$v(x_j) = v_j$$
$$v'(x_j) = \theta_j$$

Load:

$$q(x) = w_i \frac{x_j - x}{l} + w_j \frac{x - x_i}{l}$$

in which $l = x_j - x_i$. It is straightforward to integrate directly the governing equation in this case:

$$EIv(x) = \frac{w_i}{120l}(x_j - x)^5 + \frac{w_j}{120l}(x - x_i)^5 + c_1 x^3 + c_2 x^2 + c_3 x + c_4$$

The constants in this expression must be chosen to satisfy the boundary conditions listed above. This involves solving four equations for the four unknowns c_1, c_2, c_3, and c_4. This takes a bit of work, but after completing the necessary algebra we get the following solution:

31

$$v(x) = \frac{w_i}{120EIl}\left[(x_j - x)^5 + 2l^3(x_j - x)^2 - 3l^2(x_j - x)^3\right]$$

$$+ \frac{w_j}{120EIl}\left[(x - x_i)^5 + 2l^3(x - x_i)^2 - 3l^2(x - x_i)^3\right]$$

$$+ v_i N_1(x) + \theta_i N_2(x) + v_j N_3(x) + \theta_j N_4(x)$$

in which

$$N_1(x) = 1 - \frac{3(x - x_i)^2}{l^2} + \frac{2(x - x_i)^3}{l^3}$$

$$N_2(x) = x - \frac{2(x - x_i)^2}{l} + \frac{(x - x_i)^3}{l^2}$$

$$N_3(x) = \frac{3(x - x_i)^2}{l^2} - \frac{2(x - x_i)^3}{l^3}$$

$$N_4(x) = \frac{(x - x_i)^3}{l^2} - \frac{(x - x_i)^2}{l}$$

You should verify for yourself that the differential equation and boundary conditions are satisfied by making direct substitutions into the above relations.

Once we know the displacements in a beam, we can compute everything else. In particular, we can compute the slope, moment, and shear throughout the beam segment by taking the necessary derivatives. Of particular interest are the end reactions corresponding to a given load/boundary condition case, and these are given below:

$$V_i = EIv'''(x_i) = -\left[\frac{7w_i l}{20} + \frac{3w_j l}{20}\right] + \frac{12EI}{l^3}(v_i - v_j) + \frac{6EI}{l^2}(\theta_i + \theta_j)$$

$$M_i = -EIv''(x_i) = -\left[\frac{w_i l^2}{20} + \frac{w_j l^2}{30}\right] + \frac{6EI}{l^2}(v_i - v_j) + \frac{4EI}{l}\theta_i + \frac{2EI}{l}\theta_j$$

$$V_j = -EIv'''(x_j) = -\left[\frac{3w_i l}{20} + \frac{7w_j l}{20}\right] + \frac{12EI}{l^3}(v_j - v_i) - \frac{6EI}{l^2}(\theta_j + \theta_i)$$

$$M_j = EIv''(x_j) = \left[\frac{w_i l^2}{30} + \frac{w_j l^2}{20}\right] + \frac{6EI}{l^2}(v_i - v_j) + \frac{2EI}{l}\theta_i + \frac{4EI}{l}\theta_j$$

These expressions make it possible for us to calculate directly the end moments and shears associated with any specified end displacements and rotations for any beam segment with a linearly varying load. In effect these serve as constitutive relations for each beam segment, relating loads (i.e., end moments and shears) to displacements (end rotations and deflections).

Now that we have completely characterized one segment of a beam, we are ready to consider how to assemble the pieces to model a general beam system. This is most easily demonstrated by means of an example. Consider the beam shown in the figure below:

A propped cantilever with multiple loads

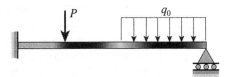

This beam can be viewed as an assemblage of three segments defined by the locations of the loads and supports:

Discretization of the beam into segments

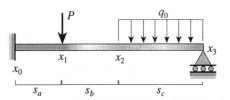

As shown in the figure, segment a (s_a) lies between the left support ($x = x_0$) and the location of the point load ($x = x_1$), segment b (s_b) lies between the point load location and the start of the distributed load ($x = x_2$), and segment c (s_c) lies between the start of the distributed load and the right support ($x = x_3$).

Using the results from above, we know we can express the displacement in each segment in terms of its respective end displacements and rotations. To accomplish this, we need to identify each segment's end data in terms of the overall system data. The following table provides a way to do this (note that you get to fill in part of the final row of the table yourself):

segment	x_i	x_j	v_i	θ_i	v_j	θ_j	w_i	w_j
s_a	x_0	x_1	0	0	v_1	θ_1	0	0
s_b	x_1	x_2	v_1	θ_1	v_2	θ_2	0	0
s_c	x_2	x_3					$-q_0$	$-q_0$

The fact the elements are connected to one another is reflected in the table by means of shared displacements, e.g., the right-end displacement and rotation of segment s_a are the same as the left-end displacement and rotation of segment s_b, and so on. Note also that some of the displacements and rotations are known to be zero from inspection by considering the support conditions. This is how we enforce the kinematic constraints of compatibility.

Now we need some means to determine the unknown displacements and rotations at the segment ends. This we accomplish by enforcing equilibrium at each inter-segment end point, or *node*. In short we require that the sum of the segment end reactions (moments and shears) at each node, plus any applied loading, balance out at the point in question. For the example above we have:

at x_1:

$$V_j(s_a) + V_i(s_b) = -P$$
$$M_j(s_a) + M_i(s_b) = 0$$

at x_2:

$$V_j(s_b) + V_i(s_c) = 0$$
$$M_j(s_b) + M_i(s_c) = 0$$

at x_3:

$$M_j(s_c) = 0$$

in which the notation $V_j(s_a)$ means the right-end shear in segment a, and so on. (Note that in the above equilibrium relations we have ignored the equations associated with known displacements and rotations known from boundary conditions. Note also that the sign convention used here is the *external* convention that is typically used in statics rather than the internal convention used in plotting moment and shear diagrams.)

These equations of equilibrium can be written in terms of the unknown displacements simply by substituting from our previous constitutive relations. Taking account of the data from the table we have the following simplified constitutive relations for each segment:

Segment *a*:

$$V_j(s_a) = \frac{12EI}{l_a^3}(v_1 - 0) - \frac{6EI}{l_a^2}(\theta_1 + 0) = \frac{12EI}{l_a^3}v_1 - \frac{6EI}{l_a^2}\theta_1$$

$$M_j(s_a) = \frac{6EI}{l_a^2}(0 - v_1) + \frac{2EI}{l_a}0 + \frac{4EI}{l_a}\theta_1 = -\frac{6EI}{l_a^2}v_1 + \frac{4EI}{l_a}\theta_1$$

Segment *b*:

$$V_i(s_b) = \frac{12EI}{l_b^3}(v_1 - v_2) - \frac{6EI}{l_b^2}(\theta_1 + \theta_2)$$

$$M_i(s_b) = \frac{6EI}{l_b^2}(v_1 - v_2) + \frac{4EI}{l_b}\theta_1 + \frac{2EI}{l_b}\theta_2$$

$$V_j(s_b) = \frac{12EI}{l_b^3}(v_2 - v_1) - \frac{6EI}{l_b^2}(\theta_2 + \theta_1)$$

$$M_j(s_b) = \frac{6EI}{l_b^2}(v_1 - v_2) + \frac{2EI}{l_b}\theta_1 + \frac{4EI}{l_b}\theta_2$$

Segment *c*:

$$V_i(s_c) = -\left[\frac{-7q_0 l_c}{20} + \frac{-3q_0 l_c}{20}\right] + \frac{12EI}{l_c^3}v_2 + \frac{6EI}{l_c^2}(\theta_2 + \theta_3)$$

$$M_i(s_c) = -\left[\frac{-q_0 l_c^2}{20} + \frac{-q_0 l_c^2}{30}\right] + \frac{6EI}{l_c^2}v_2 + \frac{4EI}{l_c}\theta_2 + \frac{2EI}{l_c}\theta_3$$

$$M_j(s_c) = \left[\frac{-q_0 l_c^2}{30} + \frac{-q_0 l_c^2}{20}\right] + \frac{6EI}{l_c^2}v_2 + \frac{2EI}{l_c}\theta_2 + \frac{4EI}{l_c}\theta_3$$

35

in which l_a = length of segment a, etc.

Substituting these constitutive expressions into the equilibrium relations above leads to the following set of equations expressed now in terms of the unknown displacements and rotations:

at x_1:

$$\frac{12EI}{l_a^3}v_1 - \frac{6EI}{l_a^2}\theta_1 + \frac{12EI}{l_b^3}(v_1 - v_2) + \frac{6EI}{l_b^2}(\theta_1 + \theta_2) = -P$$

$$-\frac{6EI}{l_a^2}v_1 + \frac{4EI}{l_a}\theta_1 + \frac{6EI}{l_b^2}(v_1 - v_2) + \frac{4EI}{l_b}\theta_1 + \frac{2EI}{l_b}\theta_2 = 0$$

at x_2:

$$\frac{12EI}{l_b^3}(v_2 - v_1) - \frac{6EI}{l_b^2}(\theta_2 + \theta_1) + \frac{12EI}{l_c^3}v_2 + \frac{6EI}{l_c^2}(\theta_2 + \theta_3) = -\frac{q_0 l_c}{2}$$

$$\frac{6EI}{l_b^2}(v_1 - v_2) + \frac{2EI}{l_b}\theta_1 + \frac{4EI}{l_b}\theta_2 + \frac{6EI}{l_c^2}v_2 + \frac{4EI}{l_c}\theta_2 + \frac{2EI}{l_c}\theta_3 = -\frac{q_0 l_c^2}{12}$$

at x_3:

$$\frac{6EI}{l_c^2}v_2 + \frac{2EI}{l_c}\theta_2 + \frac{4EI}{l_c}\theta_3 = \frac{q_0 l_c^2}{12}$$

These five linear equations are sufficient to solve for the five unknown displacements and rotations v_1, θ_1, v_2, θ_2, and θ_3. Once these displacements and rotations are known, then we can calculate everything about the beam's response from our earlier relations.

It turns out that it is not difficult to set up these system equations directly using easily automated algorithms (you can probably figure out how to do this yourself if you fiddle around with some more examples on your own). This is how Dr. Beam generates its solutions: it breaks a system into segments according to the arrangements of supports and loads, computes the various segment coefficients directly, pieces together system equations of the form shown previously (numerically rather than symbolically), and then solves these equations for the unknown end displacements and rotations. It then uses the segment relations for moments, shears, etc. to plot results as desired.

As you can see, Dr. Beam can do its work quite quickly—in fact the numerical part of the calculations for a typical case takes about 10% of the time required for each screen update. The rest of the computational time is taken up primarily in graphics.

3 ■ PHYSICAL INTERPRETATION

We can use Dr. Beam itself to get a more hands-on perspective of how this method of analysis works. In particular, we will carry out our own process of adjusting a beam's displacements at particular points to "relax" a beam into its equilibrium configuration.

■ Configuration

A simple beam with two point loads

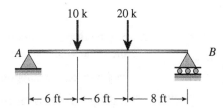

The simple beam shown above can be solved in a variety of ways, but here we will mimic the style of analysis used internally by Dr. Beam.

■ Dr. Beam File

Worksheet3a: This file contains a simple beam matching the previous configuration, except pin supports have been placed at the loading points instead of point loads. These pin supports can be used to set displacements manually at the points in question.

■ Things to Try

1. Use the pin support tool ▲ to shift-drag the two internal supports up and down. See if you can adjust the displacement of each so that the jump in the shear diagram across the left and right inner supports matches the applied 10 k and 20 k loads, respectively.

2. If you lose your patience prior to succeeding, open the files **Worksheet3b** and **Worksheet3c**. Compare these two files and you will see that everything is identical, except that in one case loads have been applied as loads, and in the other case the beam has been deformed into a particular shape.

▶ **Observation** | The solution to any beam problem can be viewed as displacing it such that the shear and moment diagrams match (i.e., are in equilibrium with) the applied loading.

3. Now let's try a more systematic approach to generating solutions this way. Return to the **Worksheet3a** case above, and set the displacement at the internal supports back to zero (or just close and open the file again).

4. Set the displacement of the left internal support (Support 1) to 1.000 in. Use the shear values to complete the following:

    required load at Support 1 = _____ k ❑ up ❑ down

    required load at Support 2 = _____ k ❑ up ❑ down

    (Remember, the required loads will match the jump in the shear values at each support—see the file **Worksheet3d** to check your results.)

▶ **Observation** | Any displacement of a beam solves *some* problem: it's just a matter of determining what loading one must apply to hold the beam in the given shape.

5. Set the displacement of the Support 1 back to zero, and then set the displacement of the right internal support (Support 2) to 1.000 in. Complete the following:

    required load at Support 1 = _____ k ❑ up ❑ down

    required load at Support 2 = _____ k ❑ up ❑ down

    (See the file **Worksheet3e** to check your results. Note that the Support 1 load due to Support 2 movement equals the Support 2 load due to Support 1 movement above. This is no accident—this type of symmetry always arises for linear structures.)

6. We now know the loads necessary to sustain unit displacements at each of the adjustable supports. Since these are linear problems, we can use these results to determine the loading required for any other displacement at these supports by simply scaling our unit results obtained above. To see this, use the file **Worksheet3e** and double-click on each load and scale by a factor of 1/10. Add a displacement label to complete the following:

   displacement at Support 1 = _____ in.

   displacement at Support 2 = _____ in.

7. What happens if both supports displace at the same time? Since we held one displacement fixed at zero while we set the other to 1 above, the forces necessary to sustain any general combination of support displacements can be determined from a simple sum of the forces required for each individually. Let's try it using the original **Worksheet3a** file. Set the displacement of Support 1 to $\delta_1 = 0.2$ in., and the displacement of Support 2 to $\delta_2 = 0.3$ in. Compare the shear jumps to the following calculated values:

   $$\text{Shear jump 1} = P_1 = \left(98.513084 \frac{k}{\text{in.}}\right)\delta_1 - \left(76.963347 \frac{k}{\text{in.}}\right)\delta_2 = \text{_____} k$$

   $$\text{Shear jump 2} = P_2 = \left(-76.963347 \frac{k}{\text{in.}}\right)\delta_1 + \left(75.42408 \frac{k}{\text{in.}}\right)\delta_2 = \text{_____} k$$

   (Note how the signs attached to the unit displacement values we determined above account for the direction of the forces.)

8. Now think back to the original question: if the loading on the beam is known, how can we determine the corresponding displacements? In other words, given P_1 and P_2, what are the corresponding δ_1 and δ_2? Based on the previous equations we need to solve the following system of equations:

   $$10\,k = \left(98.513084 \frac{k}{\text{in.}}\right)\delta_1 - \left(76.963347 \frac{k}{\text{in.}}\right)\delta_2$$

   $$20\,k = \left(-76.963347 \frac{k}{\text{in.}}\right)\delta_1 + \left(75.42408 \frac{k}{\text{in.}}\right)\delta_2$$

   Solve these two equations and two unknowns and record your results below:

   $\delta_1 = $ _____ in.

$$\delta_2 = \underline{\hspace{3cm}}\text{in.}$$

9. Use the label tool ⬚ and the **Worksheet3a** file to check your answers.

It works! ❏ True ❏ False

▶ **Observation** | By characterizing the response of a beam to unit displacements at particular reference points, we can determine the response due to loads applied at those reference points by solving a system of linear equations.

■ **Comment**

Note that this displacement-based approach did not begin with a requirement to find reactions, followed by construction of moment and shear diagrams. From an analysis point of view this means there is no need to worry whether a system is statically determinate (i.e., the reactions are easy to find), or statically indeterminate (i.e., the reactions are hard to find).

10. Open the file named **Worksheet3f** and work through the built-in tutorial to see how this problem can be solved by means of displacement and rotation manipulation.

▶ **Observation** | Dr. Beam generates solutions by adjusting reference displacements and rotations such that equilibrium is satisfied. Data go in and come out in numerical form, but the underlying solutions are exact within each segment.

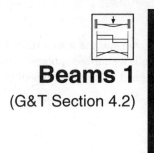

Beams 1
(G&T Section 4.2)

WHAT IF A SUPPORT DOES NOT MATCH ITS IDEALIZED CONDITIONS?

■ **Background**

As indicated in the figures below, physical beam supports do not match perfectly the idealized conditions we use in our models. Here you will investigate some of the ways this influences the accuracy of analyses based on such idealized models.

Support idealizations

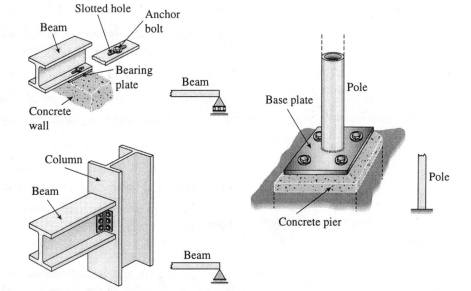

■ Dr. Beam File

Beams1a-b: These two files are identical—you can use **a** as a reference while making modifications in **b**. The configuration is a simply-supported beam with a typical loading, and labels have been pre-assigned to the quantities of interest.

■ Things to Try

1. To see what happens if the translational fixity of the right support is changed, select the linear spring tool ⬚ and click on the right support to replace the pin with a vertical spring. Compare this figure with the original pin support case.

   The displacements ❏ decrease ❏ do not change ❏ increase.

   The shears ❏ decrease ❏ do not change ❏ increase.

   The moments ❏ decrease ❏ do not change ❏ increase.

   Is this structure statically determinate? ❏ yes ❏ no

2. To see what happens if the rotational fixity of the support is changed (i.e., if we do not have a true pin), replace the right support once again with the original pin support, and use the rotational spring tool ⬚ to add a spring by clicking on the right support. This will add partial rotational constraint to the pin. Again, observe what happens and complete the following:

   The displacements ❏ decrease ❏ do not change ❏ increase.

   If you ignored the rotational constraint in design, this would be ❏ conservative ❏ unconservative in regards to calculating displacements.

   The left shear reaction ❏ decreases ❏ does not change ❏ increases.

   The right shear reaction ❏ decreases ❏ does not change ❏ increases.

   The maximum moment ❏ decreases ❏ does not change ❏ increases.

   The right end moment ❏ decreases ❏ does not change ❏ increases.

   If you ignored the rotational constraint in design, this would be ❏ conservative ❏ unconservative in regards to calculating moments.

   Is this structure statically determinate when the spring is present? ❏ yes ❏ no

3. If you double click on the rotational spring, you can adjust its stiffness by clicking and dragging in the dialog plot (Macintosh only), or by entering values in the given fields. By clicking the **Apply** button, you can update the

model for the new stiffness. Use this capability to investigate how the various quantities of interest change as you go from a small amount of constraint to a nearly fixed support.

As the rotational stiffness of the support increases, the amount of moment carried at the support ❑ decreases ❑ does not change ❑ increases.

4. Set the rotational spring stiffness back to an intermediate value and click the **Close** button in the dialog. Now let's see what happens if we add some vertical "give" at the same time we have some rotational fixity. Replace the right pin support with a linear spring again, and answer the same questions as above noting the similarities and differences:

The displacements ❑ decrease ❑ do not change ❑ increase.

The shears ❑ decrease ❑ do not change ❑ increase.

The moments ❑ decrease ❑ do not change ❑ increase.

Is this structure statically determinate? ❑ yes ❑ no

5. Adjust the rotational spring stiffness at the right support and see if you can generate an overall maximum moment that exceeds the original idealized case. Because the maximum moment is generally the most important quantity for design, this provides an evaluation of the conservative or unconservative nature of the original idealization.

I ❑ was ❑ was not able to find a spring stiffness that resulted in an overall maximum moment bigger than the original beam.

▶ **Observation** Idealizations that ignore *constraint* (like that provided by the rotational springs here) are generally conservative across the board provided one can ensure ductile behavior. Idealizations that ignore *flexibility* (like the linear springs) will not change moments and shears in statically determinate beams, but will lead to unconservative predictions of displacements.

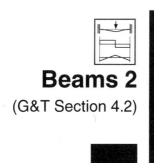

Beams 2
(G&T Section 4.2)

HOW DOES LOAD DISTRIBUTION EFFECT A BEAM'S RESPONSE?

■ Background

It is important to understand how different load types influence a beam's behavior. It is also important to realize that in real applications it is very difficult to predict what the actual physical loads will be on a given structure. If you think about something either as simple as a toothbrush or as complex as a bridge or an airplane, it is not difficult to see that it is not possible to know at the design stage exactly what loads will be applied to an object during its life. Given this uncertainty, it is important to understand in general terms how different load distributions influence a beam's response. It is also helpful to see how concentrated loads are related to distributed loads as shown below.

Basic types of loads

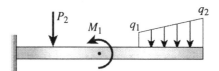

■ Dr. Beam Files

Beams2a-b: These configurations represent two limiting cases of applying a load to a simply-supported beam. In **Beams2a,** the load is concentrated entirely at the beam's center, while in **Beams2b** the same total load has been uniformly distributed over the beam's length.

■ Things to Do

1. Verify that the total load applied to the beam is the same in both cases, and compute the ratio of the concentrated load response to the distributed load response for each of the following:

    Maximum moment ratio = _____

    Maximum shear ratio = _____

    Maximum displacement ratio = _____

2. Now open file **Beams2c.** Here the same total load as described previously has been distributed at three equally spaced locations.

    The moments and displacements are more similar to the ❑ distributed load case ❑ single point load case

3. Generate your own case for five equally spaced loads, and construct plots of maximum moment and displacement as functions of the number of loads (note that the uniform load case represents an infinite number of loads).

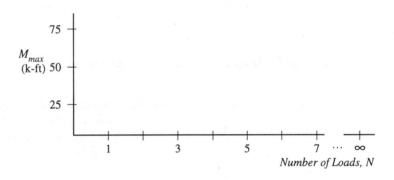

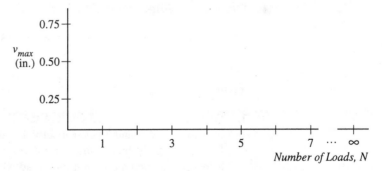

4. Return to case **Beams2a** and move the single point load from the beam's center until the maximum moment matches that of the distributed load case. Complete the following:

   x-location of load = _____ ft

   Do the maximum displacements match? ❑ yes ❑ no

5. Change the supports for cases **Beams2a** and **Beams2b** to fixed supports, and repeat the ratio computations from step **1** (note that the maximum moments occur at the supports for the fixed end case, so you will need to add additional labels):

   Maximum moment ratio = _____

   Maximum shear ratio = _____

   Maximum displacement ratio = _____

▶ **Observation** | Loads concentrated away from supports cause larger moments and displacements than loads distributed uniformly along a beam. Equally spaced loads can be reasonably well-approximated by an equivalent distributed load when the number of loads reaches about _____.

CONCENTRATED LOADS AS DISTRIBUTIONS

Concentrated loads and moments are idealizations that cannot be fully realized in the physical world. Let's look at these loadings a bit more closely to see how we can interpret them both conceptually and mathematically.

■ **Dr. Beam File**

Beams2d: A simple beam with a total load of 1k distributed over a 4 ft span.

■ **Things to Try**

1. Double click on the load and use the dialog to set the starting and ending locations to 9 ft and 11 ft respectively, and set the magnitudes to 0.5 k/ft. Push the **Apply** button, but do not close the dialog. Note that we have in effect maintained a constant total load while halving the length of beam on which it acts.

2. Repeat the process of step **1**: set the start and end locations to 9.5 ft and 10.5 ft respectively, and the load magnitudes to 1.0 k/ft. Hit **Apply** again, and observe the plots. By now you can probably guess where this is going.

3. Set the start and end locations to 9.95 ft and 10.05 ft and the magnitudes to 10 k/ft and hit **Apply** one more time. Note the similarity of the plots to a point load. If you kept going with this process, in the limit you would have an infinite load acting on a span with zero length, which clearly must be interpreted using the concept of a limit.

Having seen how a point load can be viewed as a limiting case of a distributed load, let's take a look at concentrated moments.

■ **Dr. Beam File**

Beams2e: A simple beam with two point loads forming a couple.

■ **Things to Try**

1. By increasing the loads' magnitudes while decreasing their spacing such that the net couple caused by the forces remains constant, we can obtain increasingly improved approximations of a concentrated moment. In the limit, the loads become infinite while their moment arm goes to zero, but the net moment remains the same. This is a bit tedious to do yourself so you can let Dr. Beam illustrate the process for you. Use the play button to view the process.

▶ **Observation** | Concentrated loads and moments can be viewed as limiting cases of distributions of simpler loadings.

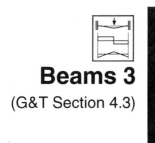

Beams 3

(G&T Section 4.3)

IS THERE A GENERAL WAY TO INTERPRET THE MOMENT SIGN CONVENTION?

■ Background

Deformation-based sign convention for moments

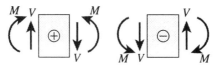

The goal here is to broaden your understanding of the deformation-based moment sign convention shown above so that you can use your physical intuition to interpret moment diagrams and their relation to beam displacements. It is also useful to have a sign convention that does not depend on the orientation of the beam or the observer. For example, in the case of the shear sign convention indicated above, note that if you walk around and look at the beam from the back, a positive shear apparently becomes a negative shear. Think also of applying the moment convention in the case a beam that is oriented vertically rather than horizontally — is the beam's "up" direction now to the left or to the right?

■ Dr. Beam File

Beams3a: A simply-supported beam with two point loads and the corresponding displacements and moments. Consistent with the sign convention for moments pictured above, the moment is positive when the top portion of the beam is in compression and negative when the bottom portion is in compression.

■ Things to Do

1. Play around with the location and magnitudes of the loads, and observe how the beam's concavity stays in sync with the moment diagram. Use your observations to complete the following:

   Where the compression is on the top portion of the beam, the moment diagram plots ❑ above (on top of) or ❑ below (beneath) the reference line.

   Where the bottom portion of the beam is in compression, the moment diagram plots ❑ above (on top of) or ❑ below (beneath) the reference line.

■ Comments

Now we can generalize the convention described above by dropping temporarily the terms "positive" and "negative" in regards to the moment. Instead, we can use a simple geometric observation: the moment diagram always plots on the compression side of the beam. This provides a useful way of reading a moment diagram: the moment diagram tells one directly where the compression side or "face" of the beam is.

There are at least two major reasons why this view of moment diagrams is useful: (i) one can use simple sketches of deflected shapes to aid in the construction and verification of moment diagrams; and (ii) one can easily construct and interpret moment diagrams for beams oriented arbitrarily in space. Let's try it out.

■ Dr. Beam Files

Beams3b-d: You probably already have a reasonably good sense of how things bend when they are loaded, and with a little practice you can get very good at qualitative prediction. If you are able to sketch deformed shapes of beams for given loadings, you can then determine by inspection where the moment diagram will be positive and negative by looking at the curvature of your sketch. This is a very useful skill for both approximate analysis and for checking results—two crucial parts of practical engineering work. You can use the files **Beams3b-d** to begin practicing.

■ Things to Do

1. Each file will open with the plots turned off. For each case, draw a sketch of the displaced shape on a piece of paper, and then predict where the moment will plot by determining the compression face of the beam. You can then

use the **Plots...** dialog in the **Options** menu to turn on the moment and displacement plots and see how you did.

2. Construct more of your own cases for practice, and challenge yourself and your friends (or your instructor) to identify where the moment diagrams plot.

3. The figure below shows a bending moment diagram for a vertically-oriented member. Add to the figure your own sketch of the displaced shape. Note how the moment diagram identifies where the compression face of the beam is without any need for thinking in terms of positive and negative bending, and as mentioned above one can easily construct and interpret moment diagrams for beams oriented arbitrarily in space.

A vertical beam

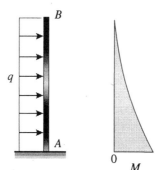

4. This would be a good time to go through Worksheet2 if you have not already done so. There you will see other ways to interpret and construct moment diagrams and learn about some alternative sign conventions. For completeness, it is worth mentioning here that in different parts of the world and among different professional communities it is common practice to plot moment diagrams on the *tension* face rather than the *compression* face of beams. If you encounter a moment diagram that appears to be upside down or backward, it may simply have been plotted with a different sign convention—check the displaced shape to be sure.

▶ **Observation** | Moment diagrams can be viewed as identifiers of the compression (or tension) side of beams.

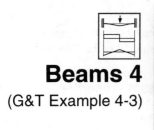

Beams 4
(G&T Example 4-3)

WHAT GENERAL PROCESS IS USED TO SOLVE PROBLEMS WITH ANALYSIS TOOLS LIKE DR. BEAM?

■ Background

Dr. Beam is designed to solve a special class of structural mechanics problems, but the basic steps and ingredients associated with setting up and solving a problem are similar to what one would need to solve any problem in structural mechanics. In this touchpoint these basic steps are illustrated in the context of a simple example for which we have an analytical solution available.

A typical beam configuration

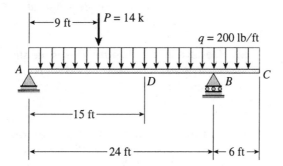

The problem above is completely specified numerically, and so Dr. Beam can be used to obtain results directly. Simple hand analysis shows that the shear and moment at point D are –6 k and 58.5 k-ft, respectively. To obtain these results using Dr. Beam, use the general solution steps that follow.

■ Solution Steps

1. **Prepare for analysis**: Open Dr. Beam. (If Dr. Beam is already open, open a new window using the **New** command in the **File** menu.)

2. **Define the basic geometry** (beam length, cross-sectional properties) **and material properties**: Use the **Beam Properties...** command in the **Info** menu to set the beam's length to 30 ft (= 24 ft + 6 ft). For a statically determinant beam we have seen that the cross-sectional properties and material properties do not influence the moments and shears, and so for this problem you can leave the default values as they are. If you are interested in displacements, though, then the appropriate material and cross-sectional properties must be entered.

3. **Define the support conditions (boundary conditions)**: Use the select tool ⌖ or the simple support tool ⟁ to double-click on the right support, and set its location to 24 ft (you can also simply drag the support to the appropriate location, but this may not be as precise in general). The left support is already located correctly for this configuration.

4. **Apply the loads**: Use the distributed load tool ⤵ to click and drag a new load. Double-clicking will open a dialog box that will allow you to enter precise data: the start and end locations are 0 ft and 30 ft respectively, while the start and end magnitudes are both 0.2 k/ft for this case. To add the point load, use the point load tool ⤓ and either click and drag, or double-click to adjust the load's magnitude and location to 14 k and 9 ft respectively.

5. **Generate a solution**: This is the step that normally takes a long time to do by hand. Dr. Beam can do this part in milliseconds, and so it is done automatically.

6. **Extract the desired values or observations from the solution**: Use the label tool ▦ to attach labels to the shear and moment diagrams at point D (located at $x = 15$ ft as shown in the figure). Simply click on each plot at the appropriate location, and double-click for precise adjustment and more significant figures in the value readout.

7. **Check and interpret your results**: In this case we have the luxury of an analytical solution to which we can compare. Check the Dr. Beam values against those given previously. If you did everything right, you should get the correct values. If something is fishy, open the file **Beams4a** and compare with your set-up.

■ Comments

With traditional methods, step **5** requires about 90% of the effort. From a design point of view, the rest of the steps require about 90% of the real engineering judgment, and so tools like Dr. Beam allow one to concentrate more effort where it is most important: on modeling, checking, and interpreting results.

8. Identify some physical beams in real world structures and use Dr. Beam to build and interpret analytical models. See how reasonable your results are.

▶ Observation

Computer-based tools can make it simple to solve structural mechanics problems quickly, but their use requires the same basic process of analysis required by any method.

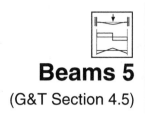

Beams 5

HOW DO SHEARS AND MOMENTS PLOT IN THE CASE OF DISTRIBUTED LOADING?

■ **Background**

The relations between shear and applied loads, and moment and shear can be expressed succinctly as follows:

$$\frac{dV}{dx} = -q \,; \quad \frac{dM}{dx} = V$$

We have seen in beam analysis that the nature of the functions describing loads, moments, and shears have relatively complex continuity properties. Kinks and jumps are common, and this requires special attention from a mathematical point of view. In this and the following two touchpoints we will take a closer look at these issues.

■ **Configuration**

A cantilever beam with a ramp load

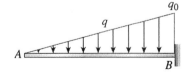

For simple cases, the derivative relation above is relatively easy to confirm. For the example shown above, the expressions for moment and shear can be determined from statics to be:

$$q = -\frac{q_0 x}{L}$$

$$V = -\frac{q_0 x^2}{2L}$$

$$M = -\frac{q_0 x^3}{6L}$$

and so clearly the relations hold in this case. Now let's look at this graphically.

■ Dr. Beam File

Beams5a: This file matches the configuration shown above, and provides a graphical view of the corresponding equations also given above.

■ Things to Do

1. Make the distributed ramp load a uniform load by double-clicking with the distributed load tool 📊 and setting the starting magnitude to match the ending magnitude. Now the relations are easier to see graphically: The shear becomes a straight line whose constant slope corresponds to the constant applied loading, and the moment becomes parabolic with a linearly varying slope consistent with the shear diagram.

2. Double-click on the load again and set the starting location to the midpoint of the beam. Note the piecewise nature of the resulting plots. Taking each portion of the beam in turn, answer the following:

    • Left of the midpoint ($x < L/2$):

    The shear diagram is ❑ constant ❑ linear ❑ parabolic and its slope is positive/zero/negative.

    The applied loading is ❑ constant ❑ linear ❑ parabolic and its sign is ❑ positive ❑ zero ❑ negative.

    The moment diagram is ❑ constant ❑ linear ❑ parabolic and its slope is ❑ positive ❑ zero ❑ negative.

    The shear diagram is ❑ constant ❑ linear ❑ parabolic and its slope is ❑ positive ❑ zero ❑ negative.

• Right of the midpoint ($x > L/2$)

The shear diagram is ❏ constant ❏ linear ❏ parabolic and its slope is ❏ positive ❏ zero ❏ negative.

The applied loading is ❏ constant ❏ linear ❏ parabolic and its sign is ❏ positive ❏ zero ❏ negative.

The moment diagram is ❏ constant ❏ linear ❏ parabolic and its slope is ❏ positive ❏ zero ❏ negative.

The shear diagram is ❏ constant ❏ linear ❏ parabolic and its slope is ❏ positive ❏ zero ❏ negative.

• At the midpoint ($x = L/2$):

The applied loading is ❏ negative ❏ positive ❏ zero ❏ undefined.

The shear is ❏ negative ❏ positive ❏ zero ❏ undefined.

The moment is ❏ negative ❏ positive ❏ zero ❏ undefined.

▶ **Observation**

The basic derivative relations given above hold within each region of loading, and at the junctions between regions abrupt changes in the loading give rise to kinks in the shear diagram.

The abrupt change of slope at the kink is qualitatively consistent with the derivative relation and the loading change. Let's see if it matches up quantitatively, too.

3. Use the **Show Values** command in the **Options** menu to turn on the numerical feedback window at the bottom of the screen. By placing (without clicking) the mouse to the right of the beam you can read off the shear at the right support. Use this value to complete the following calculations:

Shear slope to left of midpoint = _____ k/ft

Shear slope to right of midpoint = $\dfrac{\rule{1.5cm}{0.4pt}}{L/2} = \dfrac{\rule{1cm}{0.4pt}}{10\,ft} = $ _____ k/ft

Change in slope of shear diagram at midpoint = _____ k/ft

Change in applied load at midpoint = _____ k/ft

These results are all consistent with what the equations predict. ❏ yes ❏ no

4. Now use the distributed load tool ▦ to slide the load to the left so that it is located near the beam's center. Convince yourself that in each of the three regions of loading, the loading, shear, and moment diagrams obey the derivative relations and that the behavior across the load jumps is also consistent with the previous case.

5. Click and drag vertically near the left end of the distributed load until you end up with a ramp load again. Note the existence and location of kinks and jumps now.

▶ **Observation** | Abrupt changes in loading correspond to an abrupt change in the slope of the shear diagram, which results in a kinked shape. Although it is kinked, in this case the shear diagram is continuous (i.e., no jumps), and so the moment diagram has a continuous slope, i.e., no kinks.

Beams 6
(G&T Section 4.5)

HOW DO THE SHEAR-LOAD AND MOMENT-SHEAR RELATIONS PLOT IN THE CASE OF CONCENTRATED LOADING?

■ Background

We continue our examination of the load, shear, and moment relations given below:

$$\frac{dV}{dx} = -q \; ; \; \frac{dM}{dx} = V$$

Here we will consider the case of point loads.

■ Dr. Beam File

Beams6a: A simple beam with a concentrated load.

■ Things to Do

1. Based on your observation of the plots shown by Dr. Beam for the concentrated load case, answer the following questions:

   • Left of the load

   The shear diagram is ❏ constant ❏ linear ❏ parabolic and its slope is ❏ positive ❏ zero ❏ negative.

   The applied loading is ❏ constant ❏ linear ❏ parabolic and its value is ❏ positive ❏ zero ❏ negative.

The moment diagram is ❏ constant ❏ linear ❏ parabolic and its slope is ❏ positive ❏ zero ❏ negative.

• Right of the load

The shear diagram is ❏ constant ❏ linear ❏ parabolic and its slope is ❏ positive ❏ zero ❏ negative.

The applied loading is ❏ constant ❏ linear ❏ parabolic and its value is ❏ positive ❏ zero ❏ negative.

The moment diagram is ❏ constant ❏ linear ❏ parabolic and its slope is ❏ positive ❏ zero ❏ negative.

At the midpoint ($x = L/2$):

The applied loading is ❏ negative ❏ positive ❏ zero ❏ undefined.

The shear is ❏ negative ❏ positive ❏ zero ❏ undefined.

The moment is ❏ negative ❏ positive ❏ zero ❏ undefined.

Note again that these observations indicate that the prior derivative relations hold within each region of loading, and at the concentrated load location there is an abrupt change (i.e., discontinuity) in the shear diagram, and a corresponding kink in the moment diagram. The moment shear relation is thus satisfied qualitatively, and if we believe the previous equation, apparently the derivative of a discontinuous function looks like a point load.

2. Let's check the moment-shear relation quantitatively. Use the pre-labeled moment value and the feedback panel at the bottom of the screen to read off shears to complete the following calculations:

Moment diagram slope to the left of the load point =
$$\frac{M_{max}}{x_{load}} = \frac{}{6ft} = \underline{\hspace{2cm}} \text{k-ft/ft}$$

Moment diagram slope to the right of the load point =
$$-\frac{M_{max}}{L - x_{load}} = -\frac{}{14ft} = -\underline{\hspace{2cm}} \text{k·ft/ft}$$

Change in slope of moment at load point = _____ k

These results are consistent with the equations. ❏ yes ❏ no

4. Use the selection tool to drag the point load back and forth across the beam, and watch how the jump in the shear and the slope change in the moment diagram at the load point stay constant.

5. Change the supports to fixed supports and continue the dragging of the load. Note how the same behavior continues.

▶ **Observation** | Concentrated loads cause discontinuities in the shear diagram and corresponding kinks in the moment diagram.

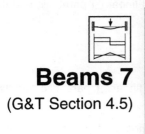

Beams 7

(G&T Section 4.5)

HOW DO THE SHEAR-LOAD AND MOMENT-SHEAR RELATIONS PLOT IN THE CASE OF CONCENTRATED MOMENTS?

■ **Background**

Here we will complete our examination of the load, shear, and moment relations as given below:

$$\frac{dV}{dx} = -q\;;\;\frac{dM}{dx} = V$$

The final case to consider is concentrated moments.

■ **Dr. Beam File**

Beams7a: A simple beam with a concentrated moment.

■ **Things to Do**

1. Based on your observation of the plots, answer the following questions:

   • Left of the load

   The applied loading is ❑ constant ❑ linear ❑ parabolic and its value is ❑ positive ❑ zero ❑ negative.

   The shear diagram is ❑ constant ❑ linear ❑ parabolic and its slope is ❑ positive ❑ zero ❑ negative.

The moment diagram is ☐ constant ☐ linear ☐ parabolic and its slope is ☐ positive ☐ zero ☐ negative.

• Right of the load

The applied loading is ☐ constant ☐ linear ☐ parabolic and its value is ☐ positive ☐ zero ☐ negative.

The shear diagram is ☐ constant ☐ linear ☐ parabolic and its slope is ☐ positive ☐ zero ☐ negative.

The moment diagram is ☐ constant ☐ linear ☐ parabolic and its slope is ☐ positive ☐ zero ☐ negative.

At the loading point ($x = L/2$):

The applied loading is ☐ negative ☐ positive ☐ zero ☐ undefined.

The shear is ☐ negative ☐ positive ☐ zero ☐ undefined.

The moment is ☐ negative ☐ positive ☐ zero ☐ undefined.

■ Comment

Note again that these observations indicate the prior derivative relations hold within each region of loading, and at the concentrated moment location there is an abrupt change (i.e., discontinuity) in the moment diagram. The discontinuities in the shear diagram at the supports correspond to the point load-like reactions there.

2. Let's check the moment-shear relation quantitatively. Use the pre-labeled moment value and the feedback panel at the bottom of the screen to read off shears to complete the following:

Change in slope of moment at load point = _____ k

Change in the moment at the load point = _____ k·ft

These results are consistent with the equations. ☐ yes ☐ no

3. Change the support conditions to those of a cantilever. Does your intuition agree with the fact there is no shear reaction at the support in this case? (You may see a small amount of plotting round-off in the shear diagram, but it is actually zero everywhere.) This is something interesting to ponder.

▶ **Observation** | Concentrated moments cause discontinuities in the moment diagram.

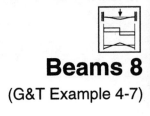

Beams 8

(G&T Example 4-7)

HOW CAN I DETERMINE MAXIMUM MOMENTS IN THE CASE OF MOVING LOADS?

■ Configuration

The problem shown below illustrates how maximum moments are associated with zero crossings of the shear diagram or concentrated moments:

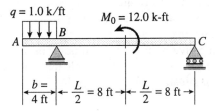

A beam with multiple loads

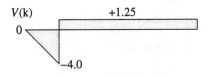

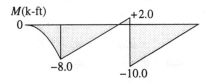

In many applications, though, the applied loads are not fixed—i.e., the locations and/or magnitudes of the loads change. One of the simplest examples is a bridge, in which the loads due to cars and trucks move across the span.

Determining maxima in these cases can be trickier than for cases of single, fixed load cases.

■ Dr. Beam File

Beams8a: An on-line version of the configuration shown previously.

■ Things to Do

1. Assume the distributed load travels across the beam and you need to determine the maximum positive and negative moments. Use the select tool [▨] to drag the distributed load along the beam, and you will see that for this case you will be able determine these maximum quantities relatively easily by trial and error.

2. A more general technique is as follows: Use the **Moment** command in the **Envelopes** menu to turn on the moment diagram envelope recorder.

3. Drag the distributed load from end to end (depending on the speed of your computer, you may need to drag slowly to get a smooth plot), and watch how the moment envelope keeps track of the maximum positive and negative moment at each point in the beam. This is a very handy tool for design as it makes it possible to see directly what and where the critical load combinations are.

4. To extract quantitative information, choose the **Inspect Envelopes** command from the **Envelopes** menu. By clicking on any of the data shown in the expanded plot view, you can read off numerical values. Use this technique to determine the maximum positive and negative moments for this case. (Note: you can increase the smoothness of the envelopes by using the **Smoothness** command in the **Envelopes** menu.)

Now let's consider another example for which the critical case is not quite as easy to determine by trial and error.

■ Dr. Beam File

Beams8b: This is a configuration somewhat similar to the previous case, except the moment load has been removed, and a point load has been added to the left end of the overhang. Also, the distributed load has been moved off the beam. Here's the new problem statement: Determine the maximum positive and negative moments on the beam assuming the distributed load can be located anywhere on the beam, and the point load may or may not be present.

■ Things to Do

1. Turn on the moment envelope using the **Moment** command in the **Envelopes** menu and then use the select tool ▶ to drag the distributed load across the beam.

2. Select the point load and type the **delete** key to remove it.

3. Move the distributed load across the beam once more, and watch how the envelope is modified for the second load case. By using the **Inspect Envelopes** command, you can extract the requested numerical results.

    Maximum positive moment = _____ k·ft

    Maximum negative moment = _____ k·ft

■ Comments

In reality, it is still not too difficult to solve this particular problem by trial and error (try it), and an experienced engineer could fairly rapidly determine the critical load combinations without too many calculations. Construction of the moment envelope, however, makes it quite simple to home in on the answer directly, and provides a method that can work even for much more complicated situations. By practicing with other simple cases, though, you should be able to improve your own ability to identify critical load cases by inspection.

■ Dr. Beam File

Beams8c: This is an envelope example that builds on some of the concepts developed in Worksheet 2: Understanding Moment Diagrams.

■ Things toTry

1. Imagine the cantilever beam shown is actually a tall tower that you are viewing sideways: its foundation is represented by the fixed support, and

the top of the building is free to move. The uniform load on the cantilever can be viewed as a simple model of a wind load blowing against the tower. The wind can of course blow in any direction, so the wind load can be imagined to cycle back and forth. Use the distributed load tool 🔲 and drag up and down on the load to model this cyclic reversal of the load (don't get carried away—keep the load between ±2.1 k/ft).

2. Now turn on the moment envelope using the **Moment Envelope** command in the **Envelopes** menu, and repeat your load cycling a time or two, trying to keep the positive and negative load extremes about the same magnitude.

   Given that moment diagrams suggest efficient structural forms, in what city was the hypothetical tower we are considering built? _____

   (Hint: see if the shape of the moment envelope looks like a particularly famous structure with which you are probably familiar.)

▶ **Observation** | Moment (or shear, or displacement) envelopes provide a useful way of keeping track of multiple load cases of the type commonly encountered in design.

Beams 9
(G&T Section 5.2)

HOW SPECIAL A CASE IS PURE BENDING?

■ Background

Pure bending is often used as a starting point for deriving beam bending equations. To get a feel for how special this case is, various configurations below have been identified as giving rise to pure bending over some or all of each beam's span. Use the indicated Dr. Beam files to investigate the fragility of these pure moment configuration. Challenge yourself to predict results correctly before having Dr. Beam show you what happens.

■ Configuration

A simple beam with end moments

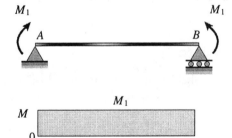

■ Dr. Beam File

Beams9a: An example of pure bending.

■ Things to Try

1. See what happens if the moments are not of equal magnitude.

2. See what happens if the moments are not at the beam's ends.

3. For a case in which the loads are located asymmetrically, see if you can adjust the moment's magnitudes to still achieve pure bending.

4. See what happens if you change the beam's length.

■ Configuration

A cantilever beam with an end moment

■ Dr. Beam File

Beams9b: An online version of the configuration shown above.

■ Things to Try

1. See what happens if the moment load moves.

2. Add a point load somewhere in the span, and see if you can adjust the moment magnitude to achieve pure moment on a portion of the span.

■ Configuration

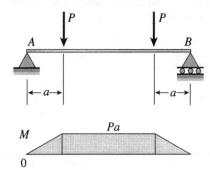

A simple beam with two point loads

■ Dr. Beam File

Beams9c: An online version of the above configuration.

■ Things to Try

1. See what happens if the point loads move.

2. See if you can find a non-symmetrically located load combination that gives pure moment on a portion of the span

■ Configuration

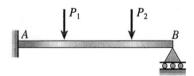

A propped cantilever

■ Dr. Beam File

Beams9d: An online version of the above configuration.

1. See if you can arrange and adjust the loads to get pure moment on a portion of the span for this case.

2. If you're still having fun, see if you can find any load and/or support arrangement involving *distributed* loads for which you can achieve pure moment. If it's not already obvious to you, see what happens for any of the cases considered here when you change the material stiffness.

▶ **Observation** | Constant or pure moment in a beam arises only for special load and support geometries and arrangements.

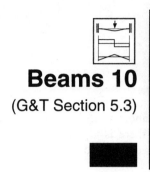

Beams 10
(G&T Section 5.3)

CAN A BEAM DEFLECT WITHOUT CURVATURE?

■ Background

Curvature is a key geometric concept in understanding beam behavior. Dr. Beam provides a helpful environment for picturing curvature in various situations, and in particular can help us answer the title question above.

■ Dr. Beam File

Beams10a: A loaded cantilever bent or curved along its entire length, i.e., with non-zero curvature throughout.

■ Things to Try

Note: If necessary, increase the deformation scale using the "up" arrow to the right of the beam 🔼, and hold the edge of a piece of paper up to the screen to see whether the plot is straight or not.

1. Move the load to the beam's midpoint and answer the following:

    The displacement to the right of the load is ❏ zero ❏ non-zero.

    The curvature to the right of the load is ❏ zero ❏ non-zero.

    The radius of curvature of the beam to the right of the load is ❏ finite ❏ infinite.

▶ Observation

This example indicates that it is possible for at least part of a beam to remain straight even when it is displaced everywhere.

■ Dr. Beam File

Beams10b: An unloaded simple beam.

■ Things to Try

1. Apply a point load acting *upwards* at about $x = 5$ ft and another at about $x = 15$ ft acting *downwards* such that there is a point of zero deflection somewhere between the loads. Now answer the following questions:

   In the vicinity of the upwards load the beam's curvature is
   ❏ concave up ❏ concave down.

   In the vicinity of the downwards load the beam's curvature is
   ❏ concave up ❏ concave down.

   At the point where the beams concavity switches, the curvature is
   ❏ zero ❏ non-zero.

   The curvature switching point (called an *inflection point*) ❏ does ❏ does not correspond to the point of zero displacement between the loads. (Before you answer, adjust the load locations and magnitudes so that the zero displacement point goes away—does the inflection point vanish, too?)

   Note: Again, increase the deformation scale and use a piece of paper as necessary to answer these questions.

2. Now delete both loads on the beam. The beam has no curvature (i.e., it is straight), but it also has no deflection. Click on the left support with the simple support tool ⛛ and drag it up and down.

   The entire beam remains straight (i.e., has zero curvature everywhere).
   ❏ true ❏ false

   The entire beam has non-zero deflection. ❏ true ❏ false

   The beam is carrying load. ❏ true ❏ false

3. See if you can cook up a load and support condition that violates the summary observation given below.

▶ Observation

It is possible for a beam to deflect without curvature, although a *loaded* beam must bend over at least a portion of its length.

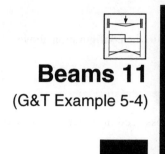

CAN REDUCED LOADS ON A BEAM INCREASE STRESSES?

■ Background

One might guess that increasing the load on a structure increases the stresses, while decreasing the load reduces the stresses. Broadly speaking this is true, but in general one must be careful as illustrated here.

■ Configuration

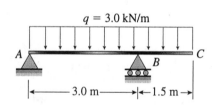

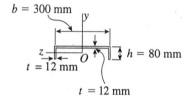

A uniformly loaded beam with an overhang and a singly-symmetric cross-section

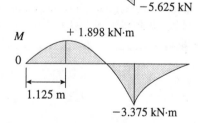

■ Dr. Beam File

Beams11a: Verify that this file corresponds to the configuration above.

■ Calculations

The top and bottom section moduli for the cross-section shown can be calculated to be:

$$S_1 = S_{top} = 133,600 \, \text{mm}^3$$

$$S_2 = S_{bottom} = 40,100 \, \text{mm}^3$$

Since we have two section moduli and a moment diagram with positive and negative moments, we need to compute the tensile and compressive stresses at each critical moment location. Using the labeled values we thus have:

$$
\begin{array}{ll}
\text{Max tension at max} & \dfrac{M}{S_2} = \dfrac{1.898 \, \text{kN} \cdot \text{m}}{40,100 \, \text{mm}^3} = 47.3 \, \text{MPa} \\
\text{positive moment} &
\end{array}
$$

$$
\begin{array}{ll}
\text{Max compression at max} & -\dfrac{M}{S_1} = -\dfrac{1.898 \, \text{kN} \cdot \text{m}}{133,600 \, \text{mm}^3} = -14.2 \, \text{MPa} \\
\text{positive moment} &
\end{array}
$$

$$
\begin{array}{ll}
\text{Max tension at max} & -\dfrac{M}{S_1} = -\dfrac{-3.375 \, \text{kN} \cdot \text{m}}{133,600 \, \text{mm}^3} = 25.3 \, \text{MPa} \\
\text{negative moment} &
\end{array}
$$

$$
\begin{array}{ll}
\text{Max compression at max} & \dfrac{M}{S_2} = \dfrac{-3.375 \, \text{kN} \cdot \text{m}}{40,100 \, \text{mm}^3} = -84.2 \, \text{MPa} \\
\text{negative moment} &
\end{array}
$$

■ Things to Do

1. Now let's remove some load from the beam and see what happens. Either double-click to get a dialog box or option-drag the right end of the load with the distributed load tool 🔲 to remove the portion of the load on the overhang.

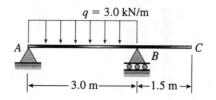

$q = 3.0$ kN/m

A

B

C

3.0 m — 1.5 m

Note how the maximum positive moment increases dramatically. Recompute the stresses using the new moment values:

Max tension at max positive moment

$$\frac{M}{S_2} = \frac{\text{kN} \cdot \text{m}}{40{,}100\,\text{mm}^3} = \underline{\hspace{2cm}}\ \text{MPa}$$

Max compression at max positive moment

$$-\frac{M}{S_1} = -\frac{\text{kN} \cdot \text{m}}{133{,}600\,\text{mm}^3} = -\underline{\hspace{2cm}}\ \text{MPa}$$

Max tension at max negative moment

$$-\frac{M}{S_1} = -\frac{-\quad\text{kN} \cdot \text{m}}{133{,}600\,\text{mm}^3} = \underline{\hspace{2cm}}\ \text{MPa}$$

Max compression at max negative moment

$$\frac{M}{S_2} = \frac{-\quad\text{kN} \cdot \text{m}}{40{,}100\,\text{mm}^3} = -\underline{\hspace{2cm}}\ \text{MPa}$$

2. Compare these new values to the previous ones and complete the following.

The maximum tensile stress ❑ increased ❑ decreased in magnitude.

■ Comment

In the event we had been dealing with a material with different strengths in tension and compression, the second, more lightly loaded case could be the critical one. This example shows how load *patterning* can be important in determining the maximum stress in a structure, and sometimes smaller total loads can be critical.

▶ **Observation** | Critical loading on a structure might not correspond to the maximum total loading on the structure.

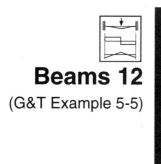

Beams 12
(G&T Example 5-5)

HOW DO DIFFERENT SUPPORT CONDITIONS EFFECT AN ACTUAL BEAM DESIGN?

■ **Configuration**

A uniformly-loaded simple wood beam

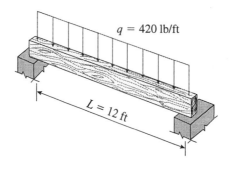

$q = 420$ lb/ft

$L = 12$ ft

The beam shown above has been designed for the case of simple supports. A summary of the design is given below:

$q = 0.42$ k/ft $+ 0.0068$ k/ft (applied load + self weight)

Allowable bending stress $= 1800$ psi

$M_{max} = 92.2$ k·in.

$S_{req} = 51.22$ in.^{3}

A suitable section is a 3×12 with a section modulus of 52.73 in.^{3}

■ **Dr. Beam File**

Beams12a: This file matches the configuration above.

■ Things to Do

1. Replace the pin supports with fixed supports. Determine the new maximum moment (be sure to check the moment at the support as well as the midspan moment).

   The maximum moment for the fixed-fixed case = _____ k·in.

2. Determine a suitable section for this new support condition. There is probably no need to worry about updating the beam's self-weight in this particular case given that it is such a small percentage of the loading, and if anything it will be reduced for the second case. (Note: you will want to refer to a table providing design properties for typical wood sections such as Appendix F in Gere & Timoshenko.)

   New section size =_____

   Reduction in area from simple support case = _____

▶ **Observation** | Fixed supports greatly reduce the demand on a beam and typically allow smaller, lighter sections to be used.

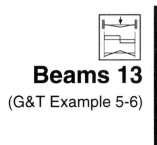

Beams 13
(G&T Example 5-6)

HOW DO STIFFNESS EFFICIENCY CONSIDERATIONS ENTER A DESIGN?

■ **Configuration**

Two proposed cantilever designs

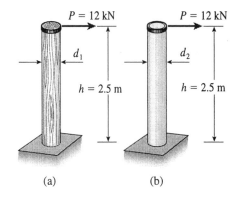

Shown above are two proposed designs for a cantilever pole: a solid wood design; and a hollow aluminum tube design. Based on allowable stresses, the following cross-sectional geometries have been selected:

Wood:

- diameter = 273 mm
- $I = 273 \times 10^6$ mm^4

Aluminum:

- outer diameter = 208 mm
- inner diameter = 156 mm
- $I = 62.8 \times 10^6$ mm^4

Given material properties of $E_{Al} = 70\,\text{GPa}$ and $E_{wood} = 11\,\text{GPa}$, let's see how the stiffnesses of the two designs compare.

■ **Dr. Beam File**

Beams13a-b: These two files are default files that must be configured for this problem. However, the units have already been set appropriately (using the **Units...** command in the **Options...** menu).

■ **Things to Do**

1. Enter the appropriate beam data, loading, and boundary conditions for the wood case in file **Beams13a**. Add a displacement label at the free end of the cantilever.

   Maximum deflection = _____ mm

2. Set up **Beams13b** for the aluminum tube case.

   Maximum deflection = _____ mm

   Ratio of deflections = _____

   Ratio of flexural rigidities, EI = _____

▶ **Observation** | Stiffness can be an important consideration in beam design, and different section and material choices play an important role in this regard.

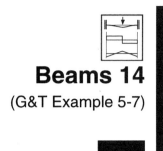

Beams 14

(G&T Example 5-7)

HOW DO I DESIGN A BEAM IF THE LOADS CAN MOVE?

■ Configuration

A steel $W12 \times 50$ has been determined to be adequate for the loading shown below assuming an allowable bending stress of 18 ksi. Will this section work if the load on the right can move? (See, e.g., Appendix E in Gere & Timoshenko to obtain section properties for a $W12 \times 50$.)

A simple beam with two distributed loads

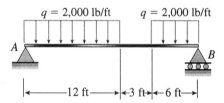

■ Dr. Beam File

Beams14a: This file matches the configuration above, including an allowance for the self-weight of the beam.

■ Things to Do

1. Turn on moment envelopes using the **Moment** command in the **Envelopes** menu.

2. Use the select tool [▶] to drag the rightmost distributed load off the right end of the beam, and then back all the way across and off the left end of the

beam (depending on the speed of your machine, you may need to drag relatively slowly to get a smooth envelope).

3. Use the **Inspect Envelopes** menu inside the **Envelopes** menu to open up an inspection window for the moment envelope.

    Maximum moment = _____ k·ft

    Location of maximum moment = _____ ft from left end

    The W12 x 50 is adequate. ❏ true ❏ false

    An adequate section = _____

▶ **Observation** | In the presence of moving loads, design generally will be governed by a critical load *location* as well as magnitude.

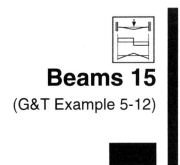

Beams 15
(G&T Example 5-12)

DO SHEAR STRESSES EVER GOVERN A DESIGN?

■ **Configuration**

A simple wood beam with two point loads

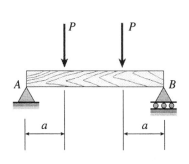

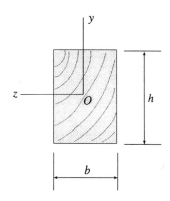

The maximum permissible values for loads shown above can be calculated to be 8.25 kN using the following:

$\sigma_{\text{allow}} = 11$ MPa

$\tau_{\text{allow}} = 1.2$ MPa

$a = 0.5$ m

$b = 100$ mm

$h = 150$ mm

The following expressions show how we can determine the maximum allowable moments and shear for this cross-section and material combination:

$$V_{\text{all}} = \frac{\tau_{all}A}{(3/2)} = \frac{\tau_{all}bh}{(3/2)} = 12 \text{ kN}$$

$$M_{allow} = S\sigma_{allow} = \frac{bh^2}{6}\sigma_{allow} = 4.12 \text{ kN} \cdot \text{m}$$

Now let's use Dr. Beam to look for load combinations that will make shear the critical limit.

■ Dr. Beam File

Beams15a: This file matches the configuration above. Displacement plotting has been turned off.

■ Things to Do

1. Note how the starting configuration is such that the maximum moment is at the maximum allowable value, while the maximum shear is less than the maximum allowable shear.

   Ignoring factors of safety, if the loads are increased, would there first be a bending failure or a shear failure? _____

2. Halve the distance of each load from the supports (i.e., use $a = 0.25$ m), and increase their magnitudes to 12 kN.

   The maximum moment is ❏ greater than ❏ equal to ❏ less than the allowable moment.

   The maximum shear is ❏ greater than ❏ equal to ❏ less than the allowable moment.

   Ignoring factors of safety, if the loads are increased, would there first be a bending failure or a shear failure? _____

3. Try other load combinations and support conditions that lead to shear-critical results. You will find that in general you need to put relatively large loads relatively close to supports.

▶ **Observation** | Normal stresses generally control the design of beams, but there are cases in which shear will govern.

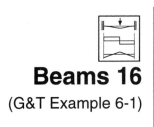

Beams 16

(G&T Example 6-1)

HOW DO I DETERMINE DISPLACEMENTS FOR BEAMS WITH COMPOSITE CROSS-SECTIONS?

■ Background

Determining the stresses in beams with composite cross-sections such as that shown below allows us to assess the strength enhancement achieved by combining materials. However, an equally important consideration for many composite designs is the effective bending stiffness of the resulting section. In this and the following touchpoint we will look at issues of bending stiffness.

A composite cross-section

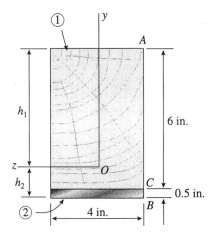

For the particular wood and steel section shown above we have for our material properties $E_1 = 1,000$ ksi and $E_2 = 20,000$ ksi. After determining that the location of the neutral axis is 5.031 in. from the top of the section, the moments of area of the cross-section components can be computed to give $I_1 = 171.0$ in.^{4} and $I_2 = 3.104$ in.^{4}. To characterize the flexural rigidity of the cross-section, we need an effective *EI* factor relating moment and the corresponding curvature. This can be calculated from any of the following relations:

$$(EI)_{eff} = E_1 I_1 + E_2 I_2 = E_1(I_1 + nI_2) = 233,080 \, \text{k} \cdot \text{in.}^2$$

in which $n = E_2/E_1$.

■ Dr. Beam File

Beams16a: A uniformly loaded, simple beam with default cross-sectional and material properties.

■ Things to Do

1. The value you see is based on Dr. Beam's default values for *E* and *I*, so we need to enter appropriate values to get the correct result for the composite section. Use the **Beam Properties...** command in the **Info** menu to open the dialog box for setting beam properties. Based on the above equations you can enter E_1 for *E* and $I_1 + nI_2$ for *I*, or any combination of *E* and *I* you wish as long as the product remains equal to the value above. Once you enter appropriate values, click **OK**, and you can then determine the correct displacements for the given cross-section.

   Maximum deflection = _____ in.

2. Imagine you are now interested in considering the relative efficiency of using the steel plates versus using either plain wood or aluminum plates. You might be tempted to simply replace the value of E_2 in the above equations with *E* for wood or aluminum, but remember that the *I*'s depend on the location of the neutral axis, which also depends on the relative material stiffnesses. You thus would need to recompute all the section properties to do your comparison. Note also that the load on the beam itself would be influenced by the beam's self-weight, and you would need to account for this, too. We will consider such comparisons in the next touchpoint in which there are not as many complicating factors.

   Something to think about for this case: if the section shown above were installed upside down, how would the section stiffness be effected? ❏ decrease ❏ no change ❏ increase

▶ **Observation** Composite sections influence a beam's stiffness as well as its strength. The stiffness can be modeled using an effective *EI* for the section.

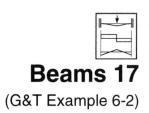

Beams 17
(G&T Example 6-2)

HOW DOES STIFFNESS DEPEND ON MATERIALS CHOICE IN A COMPOSITE CROSS-SECTION?

■ Background

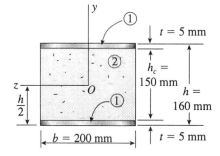

A composite cross-section

The above cross-section is doubly symmetric, so the neutral axis remains at the center of the cross-section regardless of the relative stiffnesses of the materials in question. This will make it relatively easy to examine the influence of different material combinations on bending stiffness. In particular, the geometry is such that $I_1 = 12.017 \times 10^6$ mm^4 and $I_2 = 56.250 \times 10^6$ mm^4

■ Dr. Beam File

Beams17a: For the aluminum-alloy/plastic core section shown above we have $E_1 = 72$ GPa and $E_2 = 800$MPa, and so the flexural rigidity of the section is $E_1 I_1 + E_2 I_2 = 910,200$ N $\cdot$ m^2. The relative mass/length of the composite section is given by the product of the areas and the material mass densities:

$$A_1\gamma_1 + A_2\gamma_2 = 200 \text{ mm}^2\left(28 \text{ kN/m}^3\right) + 37,500 \text{ mm}^2\left(10 \text{ kN/m}^3\right)$$

$$= 0.7 \text{ kN/m}$$

■ Things to Do

1. Increase the applied load until the maximum moment reaches the section capacity $M_p = 100$ k-ft.

   Maximum deflection = _____ mm

2. Replace the plastic core with aluminum. Enter the corresponding E and I using Dr. Beam's **Beam Properties...** command in the **Info** menu.

   Maximum deflection = _____ mm

   Beam weight = _____ kN/m

3. Replace the aluminum entirely with plastic.

   Maximum deflection = _____ mm.

   Beam weight = _____ kN/m

4. Reverse the materials in the original configuration, i.e., make the core aluminum and the outer plates plastic.

   Maximum deflection = _____ mm

   Beam weight = _____ kN/m

5. In the original configuration, the beam's self-weight represented a significant portion of the loading. Inferring that the externally applied beam load is 0.8 kN/m, complete the following table taking account of each material combination's effect on the total load:

Case	External Load (kN/m)	Self weight (kN/m)	Total Load (kN/m)	Max Moment (kN·m)	Max Deflection (mm)
Aluminum/ plastic	0.8	0.7	1.5	3.0	
Aluminum only	0.8				
Plastic only	0.8				
Plastic/ Aluminum	0.8				

► **Observation** | Prudent choices of materials in a composite section can greatly enhance a beam's overall strength and stiffness.

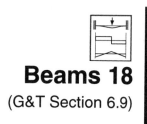

Beams 18
(G&T Section 6.9)

WHAT HAPPENS WHEN A BEAM's MAXIMUM MOMENT REACHES M_P?

■ Background

Progression of yielding in a cross-section

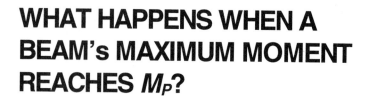

The figure above shows what happens as a beam cross-section reaches its plastic capacity in the case of elastic, perfectly plastic materials. To see the effect this has on the overall beam itself, it is necessary to consider the moment-curvature relation associated with the yielding material behavior, as shown below:

Stress strain and moment curvature in a beam.

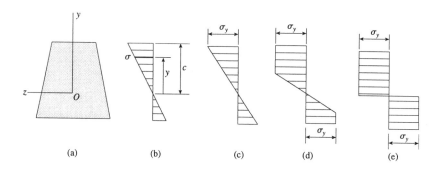

Note that the sharp yield point for the material is smoothed out for the moment curvature relation, but the general shapes of the curves are similar. The important observation relative to understanding the behavior of yielded beams is that once the beam has fully yielded, it can undergo arbitrary additional rotation without any significant increase in moment. Rotation without an increase in moment is the response associated with an internal hinge. Let's see what this looks like for a loaded beam.

■ Dr. Beam File

Beams18a: A propped cantilever beam with a point load.

■ Things to Do

1. Increase the applied load until the maximum moment reaches the section capacity $M_p = 100$ k-ft.

   The required load = _____ k

   At the fixed support the moment has reached the capacity of the section, and so a plastic hinge forms there. The following file shows how this can be modeled.

■ Dr. Beam File

Beams18b: A modified propped cantilever: the fixed support has been augmented with a hinge and a concentrated moment equal in magnitude to the capacity, M_p.

■ Things to Try

1. Compare the configurations in files **Beams18a** and **Beams18b**:

   The moment and shear diagrams for the two cases are
   ❏ identical ❏ different.

2. Increase the magnitude of the point load a small amount in case b.

   The *moment* at the fixed support ❏ decreased ❏ did not change ❏ increased.

The *rotation* at the fixed support ❑ decreased ❑ did not change ❑ increased.

This behavior is consistent with the moment-curvature plot above. ❑ true ❑ false.

3. Increase the magnitude of the point load using the up-arrow cursor key until the in-span maximum moment reaches 100 k·ft. At this point an additional plastic hinge will form at the load application point, so use the hinge tool ⊡ to add a hinge (hold down the option key while placing the hinge to get a plastic hinge).

The beam will now collapse. ❑ True ❑ False

4. Compare the magnitude of the applied load at ultimate collapse to the load necessary to cause the first hinge to form as determined above:

Collapse load = _____ k

First hinge load = _____ k

Reserve capacity beyond first hinge formation = _____ %

5. Try a simply-supported beam (or any statically determinate configuration), with a similar loading and hinge formation sequence.

Now how many plastic hinges can form prior to collapse ? _____

Reserve capacity beyond formation of first hinge = _____ %

▶ **Observation**

Once a beam reaches its plastic moment capacity at a point, it is as if a hinge with applied moments were added to the beam. An indeterminate beam typically has substantial reserve capacity beyond the formation of its first plastic hinge.

■ **Dr. Beam Note**

If you option-click with the internal hinge tool ⊡, a hinge and moments will be inserted automatically at the click location, which will model a plastic hinge assuming the moment at the point has reached M_p. (These plastic hinges will not model unloading correctly without intervention. Plasticity is a nonlinear phenomenon, and Dr. Beam is a linear program. It is possible to simulate unloading and residual states with Dr. Beam, but you will probably need help from your instructor to figure out how.)

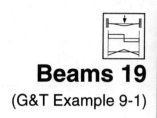

Beams 19
(G&T Example 9-1)

WHAT CAN AN ANALYTICAL SOLUTION TELL ME THAT DR. BEAM DOESN'T?

■ **Configuration**

There are many common beam configurations for which analytical solutions are relatively easy to generate, or for which analytical expression are available in tabulated form. The figure below shows one such case:

A uniformly loaded, simply supported beam

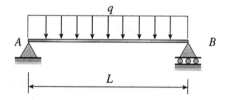

The corresponding analytical expressions for the maximum load and displacement can be written as follows:

$$M_{max} = \frac{qL^2}{8}$$

$$\delta_{max} = \frac{5qL^4}{384EI}$$

It is worth examining the value of such analytical expressions relative to direct numerical results generated by a program such as Dr. Beam.

■ Dr. Beam File

Beams19a. A uniformly-loaded, simple beam.

■ Things to Do

1. Use Dr. Beam to determine how the maximum moment and maximum deflection vary depending on the beam's length, and compare this to using the previous analytical expressions.

    This is easier with ❏ Dr. Beam ❏ the equations.

2. Determine whether the displaced shape shown by Dr. Beam is quadratic, cubic, quartic, or sinusoidal. Now try answering this question using the analytical expression below:

    $$v = -\frac{qx}{24EI}\left(L^3 - 2Lx^2 + x^3\right)$$

    This is easier with ❏ Dr. Beam ❏ the equations.

3. Investigate the change in moment and displacement if you double the load and double the moment of inertia, I. Repeat using the analytical expressions.

    This is easier with ❏ Dr. Beam ❏ the equations.

▶ **Observation** | Analytical solutions provide very useful design information at a glance.

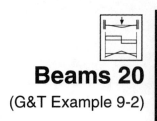

Beams 20
(G&T Example 9-2)

WHAT DOES DR. BEAM TELL ME THAT AN ANALYTICAL SOLUTION DOESN'T?

■ **Configuration**

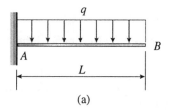

(a)

A uniformly loaded cantilever beam

Analytical expressions for moment and displacement for the cantilever shown above are given below:

$$M = -\frac{qL}{2} + qLx - \frac{qx^2}{2}$$

$$v = -\frac{qx^2}{24EI}\left(6L^2 - 4Lx + x^2\right)$$

Use these expressions to answer each of the following questions:

Where is the maximum moment located? _____

Does the displacement change sign? _____

Does the moment change sign? _____

If a 1 k load is applied on a 20 ft steel beam with $I = 140$ in.^{4}, what is the displacement 3 ft from the support? _____ in.

What will the displacement at this point be if the beam's length is increased to 30 ft? _____ in.

■ Dr. Beam File

Beams20a: A cantilever beam.

■ Things to Do

1. Answer each of the questions posed previously using Dr. Beam:

   Where is the maximum moment located? _____

   Does the displacement change sign? _____

   Does the moment change sign? _____

   If a 1 k load is applied on a 20 ft steel beam with $I = 140$ in.^{4}, what is the displacement 3 ft from the support? _____ in.

   What will the displacement at this point be if the beam's length is increased to 30 ft? _____in.

2. Once again, you be the judge:

   This is easier with ❑ Dr. Beam ❑ the equations.

▶ **Observation** | Visual and numerical presentations of results provide very useful design information at a glance.

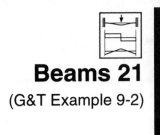

Beams 21
(G&T Example 9-2)

WHAT CAN DR. BEAM TELL ME THAT AN ANALYTICAL SOLUTION DOESN'T II?

■ **Configuration**

A point load on a beam with an overhang

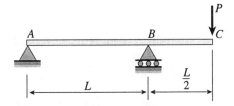

Analytical expressions for moment and displacement for the beam shown above are given below:

$$M = \begin{cases} -\dfrac{Px}{2} & ; x \le L \\ -\dfrac{P(3L-2x)}{2} & ; L \le x \le \dfrac{3L}{2} \end{cases}$$

$$v = \begin{cases} \dfrac{Px}{12EI}\left(L^2 - x^2\right) & ; \; x \le L \\ -\dfrac{P}{12EI}\left(3L^3 - 10L^2x + 9Lx^2 - 2x^3\right) & ; L \le x \le \dfrac{3L}{2} \end{cases}$$

Which of the following questions could be answered using these expressions:

Where is the maximum moment located? ❑ yes ❑ no

Does the displacement change sign? ❑ yes ❑ no

Does the moment change sign? ❑ yes ❑ no

If the support is moved to the right, how does the maximum moment change? ❑ yes ❑ no

If the point load moves to the left, how does the maximum moment change? ❑ yes ❑ no

Note that some of the important design parameters in this problem are built into the solution implicitly, and they can not be varied analytically without generating new solutions.

■ Dr. Beam File

Beams21a: This file matches the configuration above.

■ Things to Do

1. Answer each of the questions posed above using Dr. Beam:

    Where is the maximum moment located? _____

    Does the displacement change sign? ❑ yes ❑ no

    Does the moment change sign? ❑ yes ❑ no

    If the support is moved to the right, how does the maximum moment change? _____

    If the point load moves to the left, how does the maximum moment change?

    _____

▶ **Observation** | Analytical solutions generally do not allow free variation of all the parameters in a problem that are of engineering interest.

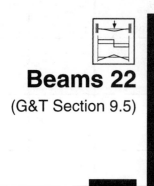

Beams 22
(G&T Section 9.5)

HOW COULD I VERIFY SUPERPOSITION EXPERIMENTALLY?

■ **Configuration**

A point load and a
uniform load on a simple
beam

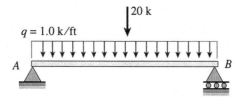

Superposition ultimately relies on the linearity of the governing equations for a beam, but it is useful to consider what an experimental verification of superposition would look like. The beam shown above is a good candidate for this exercise.

■ **Dr. Beam Files**

Beams22a, **Beams22b**, and **Beams22c**: **Beams22a** contains the case of a point load only; **Beams22b** contains the case of a distributed load only; and **Beams22c** contains the case in which both loads are acting together.

■ **Things to Do**

1. Use the label tool 🔲 to add labels (measurements in a real experiment) to the point-load only (case a) and distributed load only (case b) cases to complete the following table:

97

Case:	V_{max} (k)	M_{max} (k·ft)	θ_{max} (rads)	v_{max} (in.)	$v @ x = 4'$ (in.)
Case a: Point load only					
Case b: Distributed load only					
Case a + Case b results					
Case c: Point load and dist. load together	20	150	0.0296	2.31	1.33

Note how superposition applies to each of the quantities of interest.

2. Examine the way in which the *plots* from cases a and b add together to give case c. (If your monitor is not very big, try dragging the windows smaller.) Note how kinks and jumps combine with continuous curves.

3. It can be very useful to develop an ability to visualize how plots add together. Cook up your own superposition examples and practice.

▶ **Observation** Superposition can be used to determine any of the quantities of interest for a beam at any point within the beam.

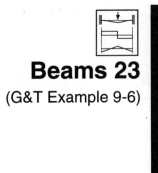

Beams 23

(G&T Example 9-6)

ARE THERE ANY OTHER INTERESTING FEATURES OF MULTIPLE LOADS BEYOND SUPERPOSITION?

■ **Configuration**

A cantilever beam with
multiple loads

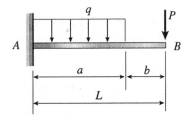

Superposition is a useful principle for generating hand solutions to problems like that shown above. Here we will discover another interesting property of beams and loads.

■ **Dr. Beam File**

Beams23a: This is a configuration similar to that shown above, but the distributed load has been replaced by a second point load.

■ **Things to Try**

1. Remove the 15 k load by selecting it with the select tool ▶ and typing the **delete** key. Record the data below:

displacement @ $x = 7' = v_{10}(7) =$ _____ in.

2. Remove the 10 k load, and then put a 15 k load back at the point $x = 7'$ (double-click with the point load tool ▼ᴾ for precise data entry). Record the value showing on the end label:

displacement @ $x = 20' = v_{15}(20) =$ _____ in.

3. Now compute the following two products:

$15\ \text{k} \times v_{10}(7) =$ _____ k·in.

$10\ \text{k} \times v_{15}(20) =$ _____ k·in.

■ Comments

products should be equal, and if you try it again with different load values and locations, you will see that this is not just a special case.

What you have just observed is the reciprocal property of linear structures. If you look at the products computed above you will see that they have units of k·in., which one normally would think of as a moment. In this case, though, these are actually the units of *work*, i.e., Force x Distance. The reciprocal property of linear structures can be stated as follows:

The work done by one set of loads applied to a structure (e.g., the 10 k load above) moving through the displacements caused by a second load set (e.g., $v_{10}(7)$) will be equal to the work done by the second load set (e.g., the 15 k load) moving through the displacements due to the first load set (e.g., $v_{15}(20)$).

There are many uses for this property, especially in the development of work and energy methods of structural analysis. Note that even for the simple case considered here, you could have algebraically computed $v_{10}(7)$ having only solved the full beam problem for the 15 k load case.

One final note: to compute work done by distributed loads requires integration, and that is why we restricted ourselves to point loads for this exercise.

Observation | There is a reciprocity among multiple loads applied individually to a beam.

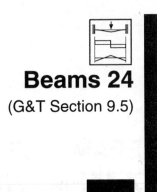

Beams 24

(G&T Section 9.5)

IS IT POSSIBLE TO HAVE ZERO MOMENT AT A POINT BUT NO INFLECTION?

■ **Configuration**

A propped cantilever beam with an overhang

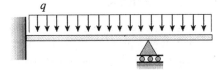

■ **Dr. Beam File**

Beams24a: This file matches the configuration above.

■ **Things to Try**

1. Look at the moment diagram, and answer the following simple questions:

   How many points are there in the beam's interior where the moment is equal to zero? _____

   How many of these points correspond to points of inflection in the beam? _____

2. Use the pin support tool ▣ to select the right pin support, and carefully drag it downwards while watching the moment diagram. See if you can make the inflection points coalesce into a single point of zero moment. Now you can answer the title question yourself.

▶ **Observation** Internal local maximum or minimum moment values can be equal to zero without associated points of inflection.

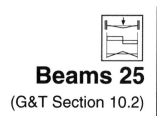

Beams 25

(G&T Section 10.2)

HOW DO STATICALLY INDETERMINATE BEAMS RELATE TO STATICALLY DETERMINATE BEAMS?

■ Background

There are two aspects to static indeterminacy that together form a "good news, bad news" pair. The good news is that from the point of view of *design*, the redundancy that makes a system indeterminate typically brings with it added reserve capacity, safety, and stiffness. The bad news is that from the perspective of *analysis*, the treatment of statically indeterminate systems often requires substantially more work. Because statically indeterminate systems tend to require more involved analysis, they are normally categorized distinctly from determinate systems on that basis alone. This categorization is somewhat artificial in that the fundamental governing equations are identical, but there are differences in *behavior* that still make distinction useful. Programs like Dr. Beam remove the work-load difference between determinate and indeterminate problems, and so one can focus on the more important practical distinctions of how indeterminate beams behave.

■ Configuration 1

The propped cantilever shown below provides the canonical example of an indeterminate beam system. Let's use Dr. Beam to see how this system works.

Propped cantilever with vertical loads

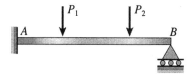

■ Dr. Beam Files

Beams25a, **Beams25b**: These two files are identical—we will use one as a reference as we make modifications in the other.

■ Things to Do

1. Place the two windows so that you can see both at the same time (you may need to resize them some).

2. Remove the right-hand pin support in the **25b** window.

    Would you say the stiffness of the beam has ❑ increased or ❑ decreased?

    Can the beam still carry load? ❑ yes ❑ no

    The beam is now ❑ determinate ❑ indeterminate.

▶ **Observation**

Because the beam does not collapse even though a support was removed, the support must have been *redundant*, i.e., not absolutely necessary to support the load.

3. Use the point load tool ⊞ to apply an upwards load at the right end of the modified, unpropped cantilever, and adjust the load's magnitude until the displacement at the end of the beam is zero (Be sure to get the load right at the end of the beam).

4. Compare the plots and values for the original and modified beams.

    The plots and values are ❑ the same ❑ different.

    The original indeterminate beam is equivalent to a determinate cantilever with an additional applied load adjusted so that the displacement boundary condition is satisfied. ❑ true ❑ false

5. Delete the newly applied load on the modified beam, and use the pin support tool ⊞ to put the right support back again. Now replace the left fixed support with a pin support.

    Would you say the stiffness of the beam has ❑ increased or ❑ decreased?

    Can the beam still carry load? ❑ yes ❑ no

    The beam is now ❑ determinate ❑ indeterminate.

6. Apply a concentrated moment at the left end of the beam and adjust its value until the rotation at the left end goes to zero.

7. Compare the plots and values for the original and modified beams.

The plots and values are ❑ the same ❑ different.

The original indeterminate beam is equivalent to a determinate simple beam with an additional applied moment adjusted so that the rotation boundary condition is satisfied. ❑ true ❑ false.

■ **Dr. Beam File**

Beams25c: This file starts with the same propped cantilever configuration and shows how one can use internal hinges to obtain a determinate beam that can be loaded to match the original indeterminate case.

▶ **Observation**
There are multiple determinate beams one can use as a starting point for mimicking indeterminate beams.

■ **Configuration 2**

Fixed-end beam with vertical loading

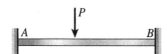

■ **Dr. Beam File**

Beams25d, **Beams25e**: These are again identical files, one for reference and one for modification.

■ **Things to Do**

1. Use the pin support tool ⌂ to replace the right fixed support with a pin support.

The stiffness of the beam has ❑ increased ❑ decreased.

Can the beam still carry load? ❑ yes ❑ no

The beam is now ❑ determinate ❑ indeterminate.

2. Now remove a second degree of redundancy by deleting the right pin support.

The stiffness of the beam has ❑ increased ❑ decreased.

Can the beam still carry load? ❑ yes ❑ no

The beam is now ❑ determinate ❑ indeterminate.

3. Try to remove one more degree of constraint by replacing the left fixed support with a pin support.

Can the beam still carry load? ❑ yes ❑ no

The beam is now ❑ determinate ❑ indeterminate ❑ unstable.

▶ **Observation**

There are degrees of indeterminacy, corresponding to how many constraints you can remove without the system becoming unstable.

4. Replace the left pin support with a fixed support again. Now we need to apply loads to the right end of the cantilever to mimic the original fixed-fixed configuration. Determine which conditions you need to satisfy to accomplish this:

❑ $v(L) = 0$ ❑ $v'(L) = 0$ ❑ $M(L) = 0$ ❑ $V(L) = 0$

5. Apply a point load and a concentrated moment at the end of the beam and adjust their magnitudes to get the end displacement and rotation to vanish simultaneously. You might find this tedious to accomplish, since adjusting two quantities to match two conditions like this is equivalent to solving a system of equations with two equations and two unknowns. Doing this by trial and error is not the best way to go. After you have tried to get it to work, see the file **Beams25f** for the solution.

The original indeterminate beam is equivalent to a determinate cantilever with an additional applied moment and point load adjusted so that the displacement and rotation boundary conditions are both satisfied.

❑ true ❑ false

■ Configuration 3

Continuous beam with
vertical loads

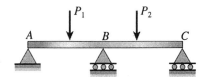

■ Dr. Beam Files

Beams25g, **Beams25h**: These are two additional identical files for
modification and comparison.

1. Carry out your own investigations in adding and removing supports and
    internal hinges, and matching indeterminate systems with adjusted loads on
    associated determinate systems.

**▶ Summary
Observations**

• One can view an indeterminate structure as one of many possible
determinate structures with redundant constraints.

• In general, the higher the degree of indeterminacy in a beam, the stiffer it is.

• The redundancy inherent in statically indeterminate systems can provide
additional reserve capacity.

• The superposition approach for solving indeterminate systems increases in
complexity with increasing degrees of indeterminacy.

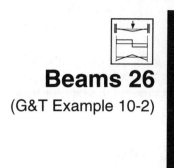

Beams 26

(G&T Example 10-2)

WHAT HAPPENS IF FIXED SUPPORTS ARE NOT TRULY FIXED?

■ **Configuration**

A fixed-end beam with a
point load at its mid-point

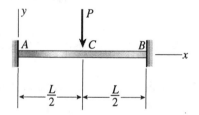

In practice it is difficult to obtain truly fixed supports, so it useful to consider
what happens if the idealized conditions are not realized.

■ **Dr. Beam Files**

Beams26a, **Beams26b**: These two files are identical—you can use one as a
reference as you make modifications in the other.

■ **Things to Do**

1. Replace the left fixed support with a pin support.

    The moment at the left support ❑ increased ❑ decreased.

    The moment at midspan ❑ increased ❑ decreased.

The displacements ☐ increased ☐ decreased.

2. Use the rotational spring tool ⊚ to apply a spring at the left support. Double-click on the new spring to make sure its location is correct, and drag the dialog box so that you can see the moment diagram for the modified beam.

3. Adjust the spring stiffness by clicking and dragging in the spring dialog stiffness plot or by entering values in the spring dialog, and use the **Apply** button to update the plots behind.

Increasing the stiffness of the spring (i.e., the fixity of the support) ☐ increases ☐ decreases the moment at the left support.

Increasing the stiffness of the spring (i.e., the fixity of the support) ☐ increases ☐ decreases the moment at the beam's center.

Increasing the stiffness of the spring (i.e., the fixity of the support) ☐ increases ☐ decreases the beam's displacements.

4. See if you can increase the stiffness enough to match the original fixed-end configuration.

The beam is now ☐ determinate ☐ indeterminate.

▶ **Observation** Overestimating the rotational stiffness of fixed supports will typically lead to an overestimate of support moments, and an underestimate of maximum in-span moments.

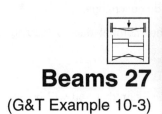

Beams 27
(G&T Example 10-3)

ARE THERE ANY SPECIAL DESIGN CONCERNS FOR STATICALLY INDETERMINATE BEAMS?

■ **Configuration**

A continuous beam with a distributed load

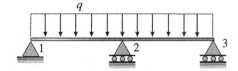

We have seen that there are design advantages associated with indeterminate beams. Here we will see one of the problematic design issues associated with indeterminate beams.

■ **Dr. Beam Files**

Beams27a, **Beams27b**: These two files match the configuration shown above and are identical—you can use one as a reference and make modifications in the other.

■ **Things to Do**

1. Use the pin support tool [⛯] to drag the center support up and down to model support settlement/misalignment.

The moments and shears ❑ change ❑ do not change.

The maximum moment can exceed the moment due to loading alone.
❑ true ❑ false

2.   Remove the right pin support. Note once again that you will be left with a
     less stiff, statically determinate beam.

3.   Use the pin support tool ▣ to again drag the center support up and down.

     The moments and shears ❑ change ❑ do not change.

▶ **Observation** | Indeterminate structures are susceptible to stresses induced by misfit, support
                      misalignment, and other load-independent effects.

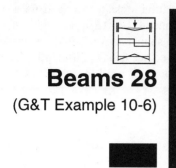

Beams 28
(G&T Example 10-6)

HOW DOES A BEAM SUSPENDED BY A CABLE BEHAVE?

■ **Configuration**

A beam supported by a cable

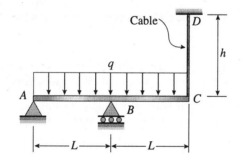

We have already considered non-rigid supports — here we will see one situation in which such supports arise.

■ **Dr. Beam File**

Beams28a: This file matches the configuration above, but without the cable.

■ **Things to Do**

1. As far as the beam is concerned, the cable looks like a simple linear spring. To determine the stiffness of the spring, consider the following relations for an axially-loaded member:

$$\delta = \frac{PL}{AE} \Rightarrow P = \frac{AE}{L}\delta \Rightarrow P = k\delta$$

This shows that the load-elongation relation looks the same as for a simple spring, with an effective spring constant of $k = AE/h$, in which h is the length of the cable for the configuration shown above. Determine the spring constant for a solid steel cable (wound cables behave differently than solid bars) with the following properties: $A = 0.1$ in.^{2}, $h = 12$ ft, and $E = 29,000$ ksi (watch your units!):

$$k = \frac{AE}{h} = \underline{\hspace{4cm}} \text{ k/ft}$$

2. Now let's add the cable to the Dr. Beam configuration. Use the translational spring tool ⟦⟧ to add a new spring to the right end of the beam. Double-click on the spring to set its location precisely, and to enter the appropriate spring stiffness from your calculation above. Drag the dialog box so you can see the plots behind, and click the **Apply** button.

    The cable ❑ increases ❑ decreases the stiffness of the beam.

3. Vary the spring stiffness between high and low values by dragging the spring dialog plot or entering values directly and clicking the **Apply** button.

    The moments and shears in the vicinity of the right end of the beam ❑ increase ❑ decrease as the cable stiffness increases.

▶ **Observation** | Axial members such as cables can be modeled as springs. Stiffer portions of structures tend to attract load.

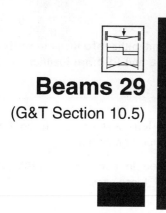

Beams 29
(G&T Section 10.5)

HOW DO NEIGHBORING SPANS INFLUENCE ONE ANOTHER IN A CONTINUOUS BEAM?

■ **Configuration**

Two continuous,
multispan beams

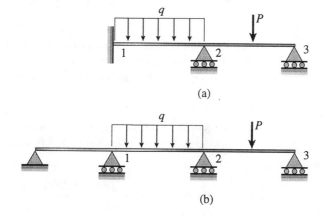

(a)

(b)

Multispan continuous beams like those shown above are common in practice, and also provide a good starting point for understanding the behavior of more complex frame structures. In this touchpoint we will consider one aspect of the interaction between neighboring spans.

■ **Dr. Beam Files**

Beams29a, Beams29b: These files match the two configurations above.

113

■ Things to Try

1. Use the **Beam Properties...** command in the **Info** menu to verify that the configurations in the files **Beams29a** and **29b** are identical on the two right-most spans.

2. Use the select tool ▣ to drag the left-most support in configuration **Beams29b** toward its neighboring support on its right.

    As the left span length is shortened in configuration **29b**, the two configurations become ☐ more similar ☐ less similar.

► **Observation** | In a continuous beam, a very short span looks to its neighboring spans like a fixed support.

3. Return the left support to its original location in configuration **29b**. Replace the fixed support at the left end of configuration **29a** with a pin support, and then add a rotational spring at the support point ($x = 0$).

4. Double-click on the spring to make sure it is located correctly, and then see if you can adjust the spring stiffness (by dragging or entering values and using the **Apply** button) so that cases **29a** and **29b** match.

5. To get an exact match, compute the following quantity and enter it as the spring stiffness: $3EI/L$ for the 5 ft span on the left end of configuration **29b**:

$$\text{rotational stiffness} = \frac{3EI}{L} = \frac{3(29000 \text{ ksi})(140 \text{ in.}^4)}{5 \text{ ft}} \times \frac{\text{ft}^2}{144 \text{ in.}^2} =$$

$$= \underline{\hspace{3cm}} \text{ k·ft/rad}$$

    In effect, the left-end span in configuration **29b** is equivalent to a rotational spring with a spring constant proportional to EI/L for the span. Note that according to this expression, as $L \rightarrow 0$, the effective spring stiffness goes to infinity, and we end up with a fixed support like you just saw above.

6. Close the spring dialog box temporarily and go once again to configuration **29b**. Replace the left-most pin support with a fixed support. Return to configuration **29a** and readjust the stiffness of the spring as before. This

time you will need a stiffness value of $4EI/L$ to get an exact match between the configurations.

▶ **Observation** | In a continuous beam neighboring spans act like rotational springs with stiffnesses proportional to EI/L.

STRESS STATES

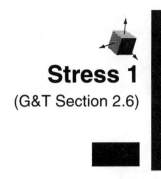

Stress 1
(G&T Section 2.6)

WHAT HAPPENS ON INCLINED PLANES IN AXIAL LOADING?

■ Background

The sequence of figures below shows a small block of material extracted from a loaded body and identifies the stress components acting on the block.

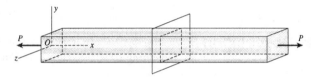

A stress block for uniaxial loading

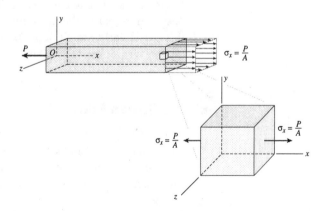

At the same location in the same body shown above, the isolation of a block of material with a different orientation leads to different components of stress:

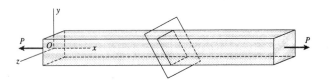

An inclined stress block for uniaxial loading

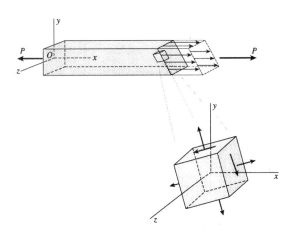

▶ **Observation**

A single state of stress at a point can give rise to different stress components depending on the orientation of the block of material we choose to consider.

It will take a while to come to a full understanding of all the ramifications of this observation, but in the meantime it is worth doing some experimentation with this particularly simple uniaxial state of stress. To this end, we will use the program Dr. Stress to help visualize the behavior of stress states.

■ **Dr. Stress File**

Stress1a: An infinitesimal block of material corresponding to the uniaxial stress state shown above in the original orientation. In this original orientation the blue face corresponds to the *x*-direction, the red face, *y*, and the green face, *z*. Stress component vectors are depicted with arrows whose arrowheads are small circles.

■ **Things to Do**

1. To get a general sense of how stress components vary as a function of material block orientation, click and drag the mouse around in the window and watch how the arrows on each face of the block change. (*Important*

Note: the stress state itself is not changing, only the material block orientation changes.)

It ☐ does ☐ does not appear that the stress arrows will keep the block in equilibrium.

2. Use the **Components** sub-menu in the **Options** menu to set the displayed components to Normal/Shear/Shear. This is the standard representation of stress, and it shows the stress component vectors on each block face decomposed into normal and shear components. Now watch what happens as you click and drag the mouse to change the block's orientation.

It is ☐ easier ☐ harder now to see whether equilibrium of the block is maintained.

3. Now let's simplify things a bit and take a 2-dimensional view. To accomplish this, first press the "o" key on your keyboard to return the block to its original (i.e., *xyz*) orientation. Use the **Components** pop-up menu in the **Options** menu to set the displayed components back to **Traction**. Use the **View** menu to select the **Snap to Green** command, which will swing the view around so that you are looking directly at the green face. (Note that in this final step, rather than changing the block orientation, you changed your view of the block.)

4. To force an in-plane rotation about the green face, simply type and hold down the "g" key on your keyboard. To rotate the other way, use "G" (i.e., shift-g).

As the block rotates the stress component arrows remain aligned with the ☐ applied load direction ☐ normals to the block faces.

5. As long as the stress component arrow (traction vector) is not normal to the block face, there will be both normal and shear stress on that face. To see this, set the components to **Shear/Normal** using the sub-menu in the **Options** menu. Rotate the block some more with the "g" key.

The shear stress arrows reach their maximum magnitude when the block is rotated _____° from the original orientation.

The maximum normal stress occurs when the block is rotated _____° from the original orientation.

6. Select the **Show Values** option in the **Options** menu to get numerical feedback of the stress values. The subscripts b, r, and g correspond to the

blue, red, and green faces, respectively, with the normal stresses denoted with σ, and the shears with τ. Continue the in-plane rotation using the "g" key to answer the following.

The maximum in-plane shear stress is ❏ 1/2 ❏ 1/3 ❏ 1/4 the maximum normal stress.

7. Go back to general 3-D block orientations by using the mouse to rotate the block (by clicking and dragging). See if you can find an orientation such that the maximum shear stress exceeds the maximum in-plane value you determined above.

I was ❏ successful ❏ unsuccessful.

8. Get back to the original orientation by typing "o" (for "original"), and then return to the original view by choosing **Original View** from the **View** menu. Set the components back to **Traction** using the **Options** menu. If you have time, repeat steps **5-7** starting with **Snap to Red** rather than green, and use the "r" key rather than "g" to rotate. See how you results compare.

Everything turns out the same. ❏ true ❏ false

▶ **Observation** | Even in the case of simple 1-dimensional, uniaxial normal stress, materials are actually subjected to combinations of normal and shear stress on planes with general orientation.

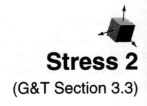

Stress 2
(G&T Section 3.3)

WHAT HAPPENS ON INCLINED PLANES IN TORSIONAL LOADING?

■ **Background**

We have seen that an apparently simple uniaxial state of stress actually induces varying amounts of normal and shear stress in a material, depending on the orientation of the material block one considers. Here we will investigate what happens in the case of pure shear, such as arises in torsional loading.

■ **Configuration**

Pure shear stress arising from torsion

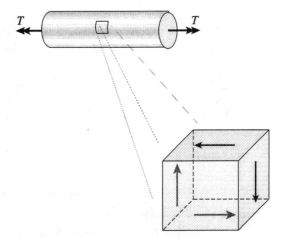

The figure above illustrates the isolation of a small block of material from a body loaded in torsion. We will once again use Dr. Stress to consider the effect of general block orientations.

■ Dr. Stress File

Stress2a: An infinitesimal block of material corresponding to the pure shear stress state shown above in the original orientation. Once again, in the initial orientation the blue face corresponds to the *x*-direction, the red face, *y*, and the green face, *z*. (Note: the **Components** option has been set to **Traction** only, so vectors are shown originating at the center of the block, which from the starting perspective makes the arrows appear offset.)

■ Things to Do

1. Let's simplify things and start with a 2-dimensional view. Use the **View** menu to select the **Snap to Green** command, which will swing the view around so that you are looking directly at the green face. (Note again that this does not change the block orientation, you are simply changing your view of the block.)

    The traction vectors are ❑ parallel ❑ perpendicular to the planes on which they act.

2. To force an in-plane rotation about the green face, simply type and hold down the "g" key on your keyboard. To rotate the other way, use "G" (i.e., shift-g). Play around a bit, and then focus on a single vector, observing what happens when you go through a full rotation of the block orientation (Note: depending on the speed of your computer you might want to speed things up increasing the sensitivity using the **Sensitivity** slider.)

    As the block rotates through a full circle, the traction vectors rotate through a ❑ half circle ❑ quarter circle ❑ full circle.

    The magnitude of the traction vectors remains unchanged. ❑ true ❑ false (If you don't trust your ability to see what's happening, select the **Show Values** option in the **Options** menu to get numerical feedback of the traction vector magnitudes. Turn the option back off when you're done.)

    Note: because compressive normal components of stress are drawn pointing towards the block's center rather than emanating from it, the traction vectors will appear to jump on occasion as the block rotates.

3. Type "o" to return to the original orientation, and then use the "g" key to determine the following:

    The angle of rotation corresponding to the case in which the traction vectors all become normal to the faces they act on = _____ °.

    The magnitude of these normal stresses are ❑ less than ❑ equal to ❑ greater than the initial shear magnitudes.

These normal stress components are ☐ tensile only ☐ compressive only ☐ tensile and compressive.

If you apply torsion to a material that is weak in tension, what would the orientation of the failure plane be? _____° (Twist a piece of chalk until it breaks to check your answer.)

4. Return the block to the original orientation with the "o" key again, and set the **Components** display in the **Options** menu to Normal/Shear/Shear. As mentioned before, this is the standard representation, and so now things will look more like the previous textbook figure. Use the "g" key to rotate some more, and watch how the normal and shear components of the stress vary. Check your result in step **3** above.

5. Use the **Show Values** option in the **Options** menu to get numerical feedback of the stress component magnitudes. (Recall that the subscripts b, r, and g correspond to the blue, red, and green faces, respectively, with the normal stresses denoted with σ, and the shears with τ.) Now use the mouse to do general 3-dimensional rotating of the block (by clicking and dragging), and see if you can find an orientation for which any stress component exceeds the starting shear magnitude.

I was ☐ successful ☐ unsuccessful.

▶ **Observation** | The pure shear arising from torsion induces compression and tension equal in magnitude to the applied shear.

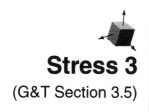

Stress 3
(G&T Section 3.5)

WHAT HAPPENS ON INCLINED PLANES WITH COMBINED TORSIONAL AND AXIAL LOADING?

■ **Background**

Having seen what happens in the cases of pure shear due to torsion and pure tension due to uniaxial loading, you might be curious about what happens if both loadings are present simultaneously. The combined case is considered here.

■ **Configuration**

A stress state arising from combined torsion and tension.

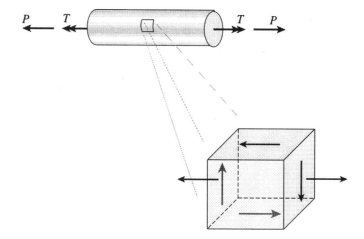

The figure above illustrates the isolation of a small block of material from a body loaded in torsion and tension. Note that the net stress state is the simple combination of the stresses from each case taken separately. Now once again use Dr. Stress to consider the effect of general block orientations.

■ Dr. Stress File

Stress3a: An infinitesimal block of material corresponding to the combined stress state shown above in the original orientation. Once again, in the initial orientation the blue face corresponds to the x-direction, the red face, y, and the green face, z. (Note: in this case the **Components** option has been set to **Normal/Shear**.)

■ Things to Do

1. Once again, start with a 2-dimensional view. Use the **View** menu to select the **Snap to Green** command to swing the view around so that you are looking directly at the green face. Use the **Options** menu to toggle the Component display back and forth between **Traction** and **Normal/Shear** to help you understand how the two depictions are related.

    The traction vector on a given face can be represented by normal and shear components. ❏ true ❏ false

2. With the **Component** display set back to **Normal/Shear**, use the "g" key to rotate the block and see if you can find an orientation for which the shear components become zero.

    I was ❏ successful ❏ unsuccessful.

3. Type "o" to return to the original orientation, and then repeat step **2** with the **Component** display set to **Traction** only. Note in this case you are seeking the orientation for which the traction vectors align with the normals of the block faces.

    I was ❏ successful ❏ unsuccessful.

    The orientation is ❏ the same ❏ different as in step 2 above.

4. It appears that for this particular stress state, it is possible to find an orientation with normal stresses only. Let's see if this is some kind of peculiar coincidence. Return the block to the original orientation with the "o" key again, and set the **Components** display in the **Options** menu

125

back to **Normal/Shear**. Choose the **Set Components** command from the **Edit** menu, and change the magnitude of the non-zero shear stress. Use the "g" key to rotate some more, and see again if you can find an orientation for which the shear components become zero.

I was ❑ successful ❑ unsuccessful.

The orientation is ❑ the same ❑ different as in steps 2 and 3 .

5. See if you can find a combination of stress components for which you are unable to find an orientation in which the shear stress vanishes. (Be sure to only set σ_{xx}, σ_{yy}, and τ_{xy}—this will keep the stress state planar. We'll consider the 3-dimensional case later.)

I was ❑ successful ❑ unsuccessful.

Note: If the size of the arrows gets too small or too big for a particular set of data, use the **Scale** slider for adjustments

■ Comment

Any plane stress state has a particular orientation for which the shear stresses vanish. This is called the *principal orientation*, and the corresponding normal stresses are called *principal stresses*. Principal stresses turn out to be very important in practical applications, and there are many additional interesting generalizations and useful observations concerning principal stresses that we will encounter later.

▶ **Observation** | The stress state arising from torsional shear and axial tension can be viewed as a simple combination of normal (tension and compression) stresses applied in a special orientation.

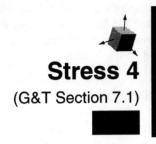

Stress 4
(G&T Section 7.1)

WHAT IS A TENSOR?

■ Background

When you first learned about velocity, you most likely began with the simple idea of speed and the basic relations between distance and time. Once you had a working understanding of these fundamental concepts as scalars, you were then introduced to the geometric aspects of velocity, i.e., the concept of velocity as a *vector*. This is similar to the process you can use to develop an understanding of stress: first you learn the basic concepts of normal and shear stresses as loads distributed over areas, and then you consider the geometric aspects of stress states. Stress states are more complex than vector quantities like velocities, and are represented by things called *tensors*. Tensor mathematics itself is relatively abstract, but fortunately there are many physical interpretations associated with stress which help ease the task of figuring it all out.

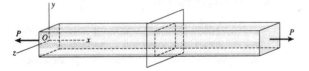

A stress block for uniaxial loading

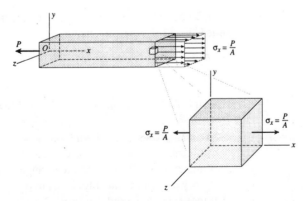

In the case of vectors, the common pictorial representation to aid understanding is simply an arrow. For stress tensors, the common pictorial representation is an

127

infinitesimal block of material imagined to be extracted from a loaded body, with the influence of the material surrounding the block represented by stress component vectors. The figure on the previous page shows the example of the generation of such a representation introduced earlier in Stress 1.

In the case of vectors, you can use an arrow-based representation to build the idea of decomposing (or projecting) a vector into scalar components relative to a particular xyz-coordinate system, e.g., $\mathbf{v} = v_x\mathbf{i} + v_y\mathbf{j} + v_z\mathbf{k}$. The tricky thing about tensors is that the block representation itself already represents a decomposition or projection of the tensor into a particular coordinate system. The vector's arrow representation provides a picture of the vector itself, while the tensor's block representation only depicts a projection—we can't actually draw a picture of the tensor itself with static 3-dimensional figures, and that's one of the main things that makes dealing with stress tricky.

Let's use Dr. Stress to see if we can extract some clarity from this apparent complexity. (Note: the tensors we will consider here represent just one class of tensors, called second-order tensors. For simplicity we will not worry about carrying along the "second-order" designation.)

■ Dr. Stress File

Stress4a: A general 3-D stress state.

■ Things to Do

1. Assuming that the components of stress on the front and back faces of a given color are equal and opposite, count up the number of component arrows on the visible faces of the block.

   There are _____ component arrows (i.e., scalar components).

   A 3-D vector has ____ scalar components, and therefore can be depicted easily in a 3-dimensional picture.

   A tensor would require a _____-dimensional picture in general.

2. Use the **Options** menu to choose the **Show Values** option. Note that only 6 component values are displayed, even though you should have noted above 3 arrows x 3 visible faces = 9 components. This is due to the symmetries in the shear stress components. Use the **Snap to Green** command in the **View** menu to change your point of view, and then use the "g" key to rotate the block around the green face. Observe the shear components on the red and blue faces.

   The shear components appear to have ❏ equal ❏ unequal magnitudes.

3. Use the **Snap to Red** command in the **View** menu to change your point of view again, and then use the "r" key to rotate the block around the red face. Observe the shear components on the green and blue faces.

The shear components appear to have ❑ equal ❑ unequal magnitudes.

4. Use the **Snap to Blue** command in the **View** menu to change your point of view again, and then use the "b" key to rotate the block around the blue face. Observe the shear components on the green and red faces.

The shear components appear to have ❑ equal ❑ unequal magnitudes.

► **Observation** | General tensors are 9-dimensional in nature (i.e., have 9 independent components), but stress tensors are symmetric in the shear components, and therefore are 6-dimensional.

5. A vector is completely characterized by its magnitude and direction, both of which are independent of how we choose to look at the vector. Let's look for some similar properties for tensors. Compute the sum of the three normal stress components for the block in its current orientation by reading the values off the screen and adding:

$$\sigma_{bb} + \sigma_{rr} + \sigma_{gg} = \underline{\hspace{3cm}}$$

6. Use the mouse (i.e., click and drag) to rotate the block to some arbitrary new orientation. Compute the sum again:

$$\sigma_{bb} + \sigma_{rr} + \sigma_{gg} = \underline{\hspace{3cm}}$$

This is ❑ equal ❑ unequal to the result above.

7. If you think this was a fluke, repeat step 6 as often as you like.

Apparently, the sum of the normal stress components is constant regardless of the block's orientation. ❑ true ❑ false.

► **Observation** | Like vectors, tensors have characterizing magnitudes and directions that do not depend on the orientation of the coordinate system used to observe the tensor. The sum of the normal components is an example of one such magnitude, and it is related to the amount of pressure contained in the stress state. Unlike a vector, a tensor has multiple characterizing magnitudes and directions rather than just one of each.

8. Turn off the values display by choosing again the **Show Values** command in the **Options** menu. See if you can find an orientation of the block such that the shear components on each face vanish. If you lose your patience, use the **Rotation** sub-menu in the **Options** menu to choose the **Principal Orientation** command.

In this orientation the stress components are purely normal.
❏ true ❏ false

9. Toggle through the various **Components** display options in the **Options** menu.

The various display options are ❏ different ❏ identical in this case.

10. Open the file **Stress4b**, which is identical to your current configuration, except the values display is turned on.

11. Bring the original window, **Stress4a**, to the front, and set the components display back to **Normal/Shear/Shear**. Choose the **Set Components** command from the **Edit** menu, and enter some new stress component values in the dialog box. (It does not matter what you enter, although if you put in very large numbers, you may need to adjust the **Scale** slider to keep everything fitting in the window. Note that the values you are entering refer back to the original *xyz*-orientation rather than the current *rgb*-orientation.)

In this orientation the stress components are still purely normal.
❏ true ❏ false.

12. Using either trial and error or the **Principal Orientation** command in the **Rotation** sub-menu in the **Options** menu, orient the block such that the shear components again vanish. Set the **Components** display option back to **Traction** only, and select the **Show Values** command in the **Options** menu.

The magnitudes of the traction vectors in **Stress4a** are ❏ equal ❏ not equal to those in **Stress4b**.

▶ **Observation**

For any stress tensor it is possible to find an orientation for which the shear stress components vanish. The corresponding normal stresses are called *principal stresses*. In general, each different stress state has its own principal orientation and principal stresses. These can be thought of as other examples of a tensor's characterizing magnitudes and directions.

13. Record the maximum traction magnitude below for either **Stress4a** or **4b** and then see if you can find an orientation of the block for which the magnitude of any traction vector exceeds this magnitude.

Initial largest principal stress magnitude = _____

I was ❑ successful ❑ not successful in finding a larger stress component magnitude by rotating the block.

■ Comment

The primary practical significance of principal stresses is that they correspond to the maximum and minimum tensile and compressive stresses acting at a point. In many cases, these maximum tensions and compressions are associated with a material's failure, and the principal orientations line up with failure planes. An example of this was mentioned earlier in Stress 2, step 3 in reference to the fracture of chalk due to torsion. If you did not try twisting a piece of chalk then, by all means try it now.

▶ Observation

Tensors can be thought of as generalizations of vectors, and they share conceptual properties and behaviors, including characterizing directions and magnitudes. Just as characterizing velocity and force requires the use of vectors, the characterizing of stress requires the use of tensors.

Stress 5
(G&T Section 7.2)

WHAT DO TRACTION VECTORS LOOK LIKE?

■ **Background**

The standard transformation equations for stress states are expressed in terms of the normal and shear scalar components acting on each face of an infinitesimally small block of material as shown below.

Transformation of stress components

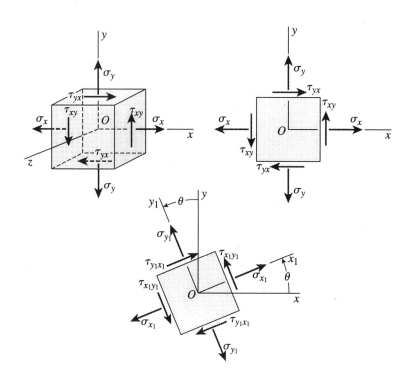

It is useful to recall that these scalar components are themselves components of traction *vectors* acting on each face, where the traction vector is defined as:

$$T_n = \lim_{\Delta A \to 0} \frac{\Delta F_n}{\Delta A}$$

Expressed in terms of traction vectors rather than scalar components, the 2-D transformation equations take a somewhat simpler form:

$$\mathbf{T}_{x_1} = (\mathbf{i}_1 \cdot \mathbf{i})\mathbf{T}_x + (\mathbf{i}_1 \cdot \mathbf{j})\mathbf{T}_y = \mathbf{T}_x \cos\theta + \mathbf{T}_y \sin\theta$$

$$\mathbf{T}_{y_1} = (\mathbf{j}_1 \cdot \mathbf{i})\mathbf{T}_x + (\mathbf{j}_1 \cdot \mathbf{j})\mathbf{T}_y = -\mathbf{T}_x \sin\theta + \mathbf{T}_y \cos\theta$$

In this touchpoint we will use Dr. Stress to observe the relatively simple behavior of traction vectors under transformation.

■ Dr. Stress File

Stress5a: A simple uniaxial stress state.

■ Things to Do

1. Let's start out with a plane perspective by swinging our view around with the **Snap to Green** command in the **View** menu. Write down the stress components you see:

   $\sigma_x =$ _____ F/L^2 (your favorite stress units go here)

   $\sigma_y =$ _____ F/L^2

   $\tau_{xy} =$ _____ F/L^2

2. Rotate the block (i.e., transform the coordinate system used to view the tensor) in the plane using the "g" key. Compare the behavior to what the transformation equations predict (i.e., look for maxes and mins and associated angles, etc.)

3. Type "o" to put the block back in its original *x-y* orientation. Now use the **Components** selector in the **Options** menu to set the display to **Traction** only.

   The display ❏ changed ❏ did not change.

   The magnitude of the traction vector equals the normal component of stress in this case. ❏ true ❏ false

   The traction vector on the x-face can be expressed simply as

   $$\mathbf{T}_x = \sigma_x \;❏\, \mathbf{i} \;❏\, \mathbf{j} \;❏\, \mathbf{k}$$

4. Use the "g" key to rotate the block again. Observe what happens and answer the following:

The traction vectors on each face ❑ do ❑ do not remain aligned with the x-direction. (Check to see if this is consistent with the traction vector transformation equation above.)

■ Comment

The behavior of the traction vectors are somewhat easier to visualize in this case than the scalar stress components themselves. The directions of the vectors remain aligned with the direction of the uniaxial loading, while their magnitudes vary with the projections of the face areas onto the uniaxial loading direction.

■ Things to Do

1. Experiment with additional stress states, toggling between the different display **Components** options to see how the different decompositions relate to one another and the particular stress state itself.

▶ **Observation** | The normal and shear components typically used to characterize stress on a plane are actually the components of a traction vector. In some cases the transformation behavior of traction vectors can be more intuitive than the individual components.

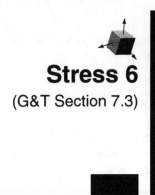

Stress 6
(G&T Section 7.3)

WHAT HAPPENS IF I ROTATE A PLANE STRESS STATE OUT OF ITS PLANE?

■ **Background**

Having gained some familiarity with plane stress analysis, you might wonder if that's the end of the story. In particular, having seen that even uniaxial stress states exhibit relatively complex behavior from a 3-D point of view, it is worth considering what happens with a plane stress state in 3D.

■ **Configuration**

A plane stress state

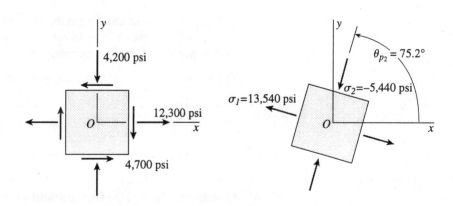

The stress state shown above is a typical plane stress example. In the following we will use Dr. Stress to observe what happens in 3D.

■ **Dr. Stress Files**

Stress6a, 6b: A plane stress state matching the configuration shown above, from two points of view.

■ **Things to Do**

1. Start with file **Stress6a** and note that the point of view for this case starts out in a planar mode. Use the "g" key to observe the in-plane rotation behavior. Verify the results indicated in the previous figure.

    The principal stresses match. ❏ true ❏ false

    The principal rotation angle looks about right. ❏ true ❏ false

    How many independent, non-zero stress components are there? _____

2. Now use file **Stress6b** to repeat step 1 from a 3-D point of view. Use only the "g" key at this point to keep the rotation planar. Note that when we draw sketches that look like the images in **Stress6a**, they are really projections of the configurations shown in **Stress6b**.

    The principal stresses match. ❏ true ❏ false

    The principal rotation angle looks about right. ❏ true ❏ false

    How many independent, non-zero stress components are there? _____

3. In both windows **6a** and **6b** put the block back in its original orientation by typing "o." Use the "r" key to rotate the orientation about the red face in each window, and observe the behavior.

    Which normal and shear stress components change?

    _____

    Which normal and shear stress components do not change?

    _____

4. In both windows use the **Components** sub-menu in the **Options** menu to set the display to **Normal/Shear**. Repeat the red rotations of step 3.

    Which normal and shear stress components change?

    _____

    Which normal and shear stress components do not change?

    _____

5. In both windows use the **Components** sub-menu in the **Options** menu to set the display to **Traction**. Once again repeat the red rotations of step 3.

    Which traction vectors change?

    _____

    Which traction vectors do not change?

    _____

6. Type "o" to put the blocks back in their original orientations in each window, and then use the "b" key to rotate about the blue face.

    Which traction vectors change?

    _____

    Which traction vectors do not change?

    _____

7. In both windows use the **Components** sub-menu in the **Options** menu to set the display back to **Normal/Shear/Shear**. Now use the mouse to generate general 3-D rotations.

    Which stress components change?

    _____

    Which stress components do not change?

    _____

    How many independent, non-zero stress components are there? _____

■ Comment

Analysis of plane stress states is useful, but it is important to remember that stress in real materials is always 3-dimensional in nature. This means that the set of normal and shear stresses experienced by a material always includes the full set of 6 components. We will see in Stress 8, that in some cases, these out-of-plane components can be more critical than the in-plane ones.

▶ **Observation** | Stress states are always embedded in 3-D space, and materials are always subject to general 3-D stress effects.

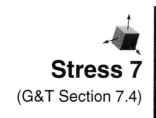

Stress 7
(G&T Section 7.4)

HOW DOES MOHR'S CIRCLE RELATE TO STRESS BLOCK DEPICTIONS?

■ **Background**

Both Mohr's circles and stress block figures are common ways to represent stress states. Here we will take a quick visual look at how these two depictions are related.

■ **Dr. Stress File**

Stress7a: A uniaxial stress state. The **Show Mohr** option has been selected in the **Options** menu.

■ **Things to Do**

1. Use the "g" key to rotate the block about the green face, and observe how the blue and red dots traverse the circle.

   The dots follow the circle. ❑ true ❑ false

2. Type "o" to return the block to its original orientation, and then choose **Snap to Green** from the **View** menu to take a planar view of the stress block. Use the "g" key to rotate the block again, and answer the following based on your observations.

   For a given dot to traverse 90° on the circle, the block rotates ____°.

   For a given dot to traverse 180° on the circle, the block rotates ____°.

3. Note that Dr. Stress only shows the dots in the upper part of the circle. This is because Dr. Stress thinks of everything in 3-D terms, and in the 3-D case Mohr's circle only carries information about the shear stress magnitudes (it's only in the 2-D case that the shear component of the traction vector is guaranteed to line up with a coordinate axis).

    For the sign convention you have learned in class, when the block rotates counterclockwise (achieved using shift-g "G"), the dots should traverse the circle ☐ clockwise ☐ counterclockwise.

■ Comment

Note that since a Mohr's circle plot shows the complete path the stress state dots will traverse for arbitrary in-plane rotation, the plot completely summarizes the stress state. At a glance one can extract the maximum and minimum stress values (principal stresses), as well as the maximum in-plane shear stress.

■ Things to Do

1. Use the **Set Components** command in the **Edit** menu to enter a more general 2-D stress state (i.e., $\sigma_{xx} \neq 0, \sigma_{yy} \neq 0, \tau_{xy} \neq 0;\ \sigma_{zz} = \tau_{xz} = \tau_{yz} = 0$). Watch what happens to the resulting Mohr's circle plot. (This phenomenon will be revisited in Stress 8.)

    I now see ___ circles instead of 1.

► **Observation**   Mohr's circles provide useful qualitative as well as quantitative information about a stress state.

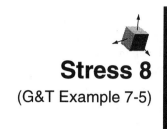

Stress 8
(G&T Example 7-5)

ARE THERE LIMITS TO PLANE STRESS ANALYSIS?

■ Background

It is straightforward to determine principal stresses and directions for plane stress states using simple equations or Mohr's circle. It is important to realize, however, that these in-plane results tell only a part of the story. In this touchpoint we will consider some important aspects of out-of-plane behavior.

■ Dr. Stress File

Stress8a: A plane stress state.

■ Things to Do

1. Use the "g" key to rotate the block into its principal orientation (don't worry about getting the shear component values to go exactly to zero—just get them close)

    To determine the maximum in-plane shear stresses, you would need to rotate the block about the green face _____° from the principal orientation.

2. Use the "g" key to determine the approximate, maximum shear stress value (you might also want to do a quick hand calculation for practice)

    The maximum in-plane shear stress is _____ F/L²

3. Return the block to the principal orientation, and then use the "r" key to rotate the block around the red face. Observe the shear stress components

on the blue and green faces as the block rotates. Determine the maximum value of these shear components.

The maximum blue-green shear stress is _____ F/L²

4. Rotate the block once again to the principal orientation, and then use the "b" key to rotate around the blue face. Observe the shear stress components on the red and green faces as the block rotates. Determine the maximum value of these shear components.

The maximum red-green shear stress is_____ F/L²

5. Compare your results from steps 2, 3, and 4 to answer the following question:

The original maximum in-plane shear stresses (see step 2) turned out to be the overall maximum shear stresses. ❑ true ❑ false

► **Observation** | Maximum shear stresses determined from a 2D analysis will not necessarily correspond to the true maximum experienced by the material.

6. Rotate the block back to the principal orientation, and then use the **Snap to Green** command in the **View** menu to take a 2-dimensional look at the problem. Rotate the block using the "g" key once more, and demonstrate to yourself that the maximum shear stress for this rotation corresponds to one-half the difference of the principal normal stress components.

$\sigma_1 =$ _____ F/L²

$\sigma_2 =$ _____ F/L²

$\left| \dfrac{\sigma_1 - \sigma_2}{2} \right| =$ _____ F/L²

7. Rotate the block to the principal orientation, and then choose the **Snap to Red** command from the **View** menu. Note that from this perspective, the stress state again appears to be 2-D, and the block is of course already in a principal orientation. Convince yourself that for this case, too, the maximum shear stress obtained by rotating in the red plane (using the "r" key) is simply half the difference of the principal normal components, σ_{bb} and σ_{gg}.

$\sigma_{bb} =$ _____ F/L²

$\sigma_{gg} =$ _____ F/L²

$$\tau_{bg\,\text{max}} = \underline{\hspace{2cm}} \; F/L^2$$

8. Rotate the block back to the principal orientation, and then choose the **Snap to Blue** command form the **View** menu. Once again, the stress state appears to be 2-D. Show that the maximum shear stress obtained by rotating in the blue plane (using the "b" key) is simply half the difference of the principal normal components, σ_{rr} and σ_{gg}.

$$\sigma_{rr} = \underline{\hspace{1.5cm}} \; F/L^2$$

$$\sigma_{gg} = \underline{\hspace{1.5cm}} \; F/L^2$$

$$\tau_{rg\,\text{max}} = \underline{\hspace{2cm}} \; F/L^2$$

▶ **Observation** | A plane stress state gives rise to three distinct planes of rotation in 3-D, each of which behaves like a 2-D stress state.

9. Return the block to its original orientation by typing the "o" key, and then return the view to the original perspective using the **Original View** command in the **View** menu. Now use the **Set Components** command in the **Edit** menu to enter an arbitrary 3D stress state.

10. Rotate the block to its principal orientation by choosing the **Principal Orientation** command in the **Rotation** sub-menu of the **Options** menu. (If you like, you can try to achieve this by trial and error.)

11. By successively snapping to green, red, and blue perspectives and performing the manipulations outlined in steps 6-8 above, see if you agree with the following statement.

    The observation above can be generalized: any stress state can be viewed as a combination of three planar stress states whose planes correspond to the block faces in the principal orientation. ❏ true ❏ false

12. Choose the **Show Mohr** option from the **Options** menu, and note that three circles appear. Each circle corresponds to one of the planar stress views mentioned above.

13. Make sure the block is in its principal orientation, and then type and hold down the "r" key to rotate about the red face. Note how the green and blue dots representing the normal/shear components on these faces traverse one

of the circles. Repeat this for the green and blue faces, being sure to return the block to its principal orientation before changing rotation colors.

The overall maximum shear stress corresponds to the smallest diameter circle. ❑ true ❑ false

14. Now use the mouse to drag and rotate the block about arbitrary axes. Watch how the dots move among the circles and see if you can achieve either of the following:

I ❑ was ❑ was not able to make any of the dots move inside the smaller circles.

I ❑ was ❑ was not able to make any of the dots move outside the largest circle.

15. Use the **Set Components** command in the **Edit** menu to enter various stress states and observe the resulting Mohr's circles. Note how the dots move among the circles for various rotations. Try the following special cases in particular: (i) uniaxial stress; (ii) pure shear; and (iii) pure tension ($\sigma_{xx} = \sigma_{yy} = \sigma_{zz} \neq 0$; $\tau_{xy} = \tau_{xz} = \tau_{yz} = 0$).

I got the biggest circles for the case of pure tension. ❑ true ❑ false

▶ **Observation**

A complete summary of a stress state can be achieved by constructing a Mohr diagram with three circles in general. These circles provide bounds on the normal and shear components for arbitrary 3-D rotation.

Stress 9

(G&T Section 7.6)

ARE THERE ANY HANDY WAYS OF THINKING ABOUT TRIAXIAL STRESS STATES?

■ **Background**

The three Mohr's circles that correspond to stress states in 3D provide a useful summary of the main quantities of interest. In this touchpoint we will consider additional ways to interpret this summary information.

■ **Configuration**

A triaxial stress state

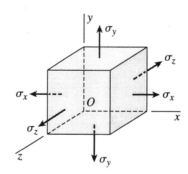

■ **Dr. Stress File**

Stress9a: A triaxial stress state corresponding to the configuration above. Note in this case that the block is already in its principal orientation.

■ Things to Do

1. Dr. Stress allows you to alter a stress state by directly manipulating the principal stress values. To see how this works, click and drag any of the stress state dots (green, red, or blue) on the Mohr's circle plot.

   When I drag the intermediate principal stress, I ☐ do ☐ do not influence the overall maximum shear stress.

   When I drag the maximum principal stress, I ☐ do ☐ do not influence the overall maximum shear stress.

   Why is it not possible to drag the principal stresses vertically? _____
   _____

2. It is also possible to drag multiple principal stress values simultaneously. Hold down the shift key while clicking on each of the three stress dots in sequence, and you will be able to drag them all together. Drag the three points back and forth, and observe the corresponding stress block behavior.

   The normal stress components increase or decrease by an equal amount.
   ☐ true ☐ false

3. Rotate the block using the mouse to some general orientation, and then repeat the shift-clicking and dragging operation for the principal stresses (note you cannot drag the colored dots once they move from the principal values).

   The shear stress components ☐ do ☐ do not change.

   The normal stress components ☐ do ☐ do not change.

■ Comment

Sliding the Mohr's circle along the normal stress axis corresponds to adding or subtracting a constant normal stress in all directions. Thus the horizontal location of the Mohr's circles indicates the amount of pure pressure or spherical tension contained in the stress state. For materials whose stiffness properties are the same in all directions (*isotropic* materials), this part of the stress state causes volume changes only. The horizontal location of the circles is quantified in terms of the average normal stress, i.e., $(\sigma_1 + \sigma_2 + \sigma_3)/3$.

4. Return the block to its original (principal in this case) orientation by typing "o". Shift-click all three principal values and drag them until their average is about zero, i.e., $(\sigma_{bb} + \sigma_{rr} + \sigma_{gg})/3 = 0$.

The volume changing component of the stress state is now zero.
❑ true ❑ false

■ **Comment**

The part of the stress that is left over after the volume-changing component has been extracted is called the *deviatoric* stress, and it is this part of the stress that has the tendency to change the shape of the material. For materials like metals that tend to yield by shearing mechanisms, it is the deviatoric stress component that causes failure. Generally speaking, the sizes of the circles' diameters characterize the magnitude of the shape-changing stress. Note that while Dr. Stress does allow you to alter the volume-changing part of the stress state by simply dragging all three principal stresses simultaneously, it does not allow you to modify independently the deviatoric portion of the stress state using simple dragging.

▶ **Observation**

Stress states can be viewed in terms of a pressure component and a shape-changing component.

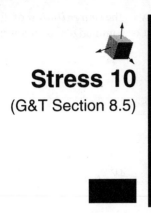

Stress 10

(G&T Section 8.5)

WHAT KINDS OF COMBINED LOADINGS CAUSE TRIAXIAL STRESS STATES?

■ **Background**

Triaxial stress states are important in many applications, and they arise naturally in regions localized around geometric, material, and loading irregularities. The types of analysis necessary to calculate such 3-D stress states, however, typically require more advanced analysis techniques than what you are likely to have encountered to date. In this touchpoint we will consider a relatively simple case that requires triaxial consideration.

■ **Configuration**

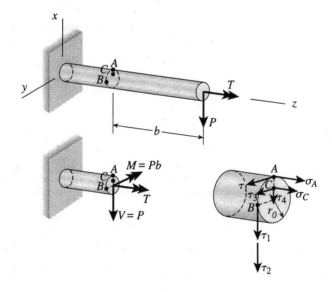

Cantilever bar subjected to combined torsion and bending

For the body loaded as previously shown, the *magnitudes* of the stress components induced by the applied loads can be summarized as follows:

- Torsional shear stress: $\tau = \dfrac{Tr}{I_p} = \dfrac{2Tr}{\pi r_0^{\,4}}$

- Normal stress due to bending: $\sigma_A = \dfrac{My}{I} = \dfrac{4My}{\pi r_0^{\,4}}$

- Shear stress due to bending:

$$\tau_{bend} = \frac{VQ}{It} = \frac{V}{\left(\pi r_0^{\,4}/4\right)} \frac{(2/3)\left(r_0^2 - y^2\right)^{3/2}}{2\sqrt{r_0^2 - y^2}} = \frac{4V}{3\pi r_0^{\,4}}\left(r_0^2 - y^2\right)$$

For this problem, the magnitudes of the various quantities are: $T = 600$ in·k; $r_0 = 2$ in; $V = 50$ k; and $M = 300$ in·k.

■ Dr. Stress Files

Stress10a-c: Blank stress states for you to fill in.

■ Things to Do

1. The configuration above identifies three points of interest, A, B, and C. Use the relations and quantities given above to determine the net stress states at each of these points.

    Point A:

    $\sigma_x =$ _____ psi $\tau_{xy} =$ _____ psi

    $\sigma_y =$ _____ psi $\tau_{xz} =$ _____ psi

    $\sigma_z =$ _____ psi $\tau_{yz} =$ _____ psi

    Point B:

    $\sigma_x =$ _____ psi $\tau_{xy} =$ _____ psi

    $\sigma_y =$ _____ psi $\tau_{xz} =$ _____ psi

    $\sigma_z =$ _____ psi $\tau_{yz} =$ _____ psi

    Point C:

    $\sigma_x =$ _____ psi $\tau_{xy} =$ _____ psi

    $\sigma_y =$ _____ psi $\tau_{xz} =$ _____ psi

    $\sigma_z =$ _____ psi $\tau_{yz} =$ _____ psi

2. Use the **Set Components** command in the **Edit** menu to enter the previous values in each of the three Dr. Stress windows corresponding to **Stress10a**, **10b**, and **10c,** respectively.

3. Use the Mohr's circles shown in each case to answer the following:

    Number of non-zero principal stresses for point A = _____

    Number of non-zero principal stresses for point B = _____

    Number of non-zero principal stresses for point C = _____

4. See if you can change the magnitudes of the non-zero stress components for point C such that all three principal stress components are non-zero.

    I ☐ was ☐ was not successful.

5. Use the "r," "g," and "b" keys to rotate the block into its principal orientation. This is most easily accomplished by observing the motion of the red, green, and blue dots on the Mohr's circle diagram, and choosing the rotation color based on the dot with the smallest shear component. (Remember that holding down the shift key reverses the rotation.)

▶ **Observation**   Interior points in bodies subjected to combined loading are likely to experience 3-dimensional stress.

Acknowledgments

This material is based on work performed at the University of Washington funded by the National Science Foundation under the auspices of the Engineering Coalition for Excellence in Education and Leadership (ECSEL). The support of each of these entities is gratefully acknowledged. Additional support was provided by the Keck Foundation.

Formal reviews were provided by the following individuals: Mr. William H. Blackmon, Carnegie Mellon University; Dr. John Dempsey, Clarkson University; Dr. Janak Dave, University of Cincinnati; Dr. Xiaomin Deng, University of South Carolina; Dr. Mohammed Elgaaly, Drexel University; Dr. Kurt Gramoll, Georgia Institute of Technology These reviews helped improve the quality of the work—responsibility for any remaining weaknesses of course resides solely with the authors.

As a hybrid of print and digital media, this product required the traversal of a significant learning curve for all parties involved. The authors would like to especially thank Gerald Barnett from the University of Washington Office of Technology Transfer for his help and guidance in this regard.

On a personal level, the completion of this project required a great deal of patience from the authors' families, and this patience has been greatly appreciated. The authors also look forward to resuming somewhat normal sleep habits.